实践景观
in practice

territoires

大地景观事务所 设计作品专辑

Green Vision 绿色观点·景观设计师作品系列

本系列图书为法国亦西文化公司(ICI Consultants/ICI Interface)的原创作品，原版为法英文双语版。
This series of works is created by ICI Consultants/ICI Interface, in an original French-English bilingual version.

法国亦西文化 ICI Consultants 策划编辑

总企划 Direction：简嘉玲 Chia-Ling CHIEN
协调编辑 Editorial Coordination：尼古拉·布里左 Nicolas BRIZAULT
英文翻译 English Translation：佐伊·迪德里奇 Zoé DIDERICH
中文翻译 Chinese Translation：简嘉玲 Chia-Ling CHIEN
版式设计 Graphic Design：维建·诺黑 Wijane NOREE
排版 Layout：卡琳·德拉梅宗 Karine de La MAISON

green vision 绿色观点·景观设计师作品系列

实践景观
in practice

territoires

philippe convercey, franck mathé, étienne voiriot

大地景观事务所 设计作品专辑

菲利普·康维尔谢、法兰克·马碟、艾提安·瓦里欧

辽宁科学技术出版社

大地景观事务所（TERRITOIRES）将基地环境的阅读推至简明的极致……此乃我所认同的一种信念。

鲁迪·里乔蒂
2013年11月

"TERRITOIRES pushes the reading of the context to the borderline of accessibility... This is a belief that I share."

Rudy Ricciotti
November 2013

GREETINGS from BOSTON MASS.
ARCHITECTURE
LANDSCAPE ARCHITECTURE
LESS IS MORE
11 MAI 2014
LE PETIT JOURNAL DE LA CITÉ
CITÉ
RICCIOTTI ARCHITECTE

献给奥利维尔 To Olivier

contents

目录

introduction

前言

我们的文化

我们的每一个景观方案皆由实践经验所主导。

大地景观事务所（TERRITOIRES）总是不断地在方案构思中寻找惊喜与意外，然而其设计结果必须是来自对空间使用者的妥善考量。这个“不期而遇”的经验，对事务所而言犹如家常便饭，使其成为各种不同尺度基地的实验室，能够在遥远而不预期之处——在其他地方并且用其他方法——为每一个场所空间所面临的特殊课题找到适当的解答。为了赋予每个方案细节处理的正当性，用心去理解每个项目组成元素的复杂性质便成为最佳的达成方法，使得设计的每一部分都能经得起现实的考验，合乎世界的某种明显常理。

大地景观事务所十分着重项目与文化的依存关系，基地环境的文化不仅塑造出各种景观样貌，也反映出居民的智慧和创造力。这份理念借助一种简约精确的明显美学而表达出来。对大地景观事务所而言，形式的存在必须能够捕捉视野、强化曲线、呈现一道光线、一段历史……形式本身不是方案的目的，而是用来诠释一个场所之需求的工具。此种审慎合宜的形式能够彰显事物的本质，使得土地和其自身拥有的特色能够自我表达，让“已经存在”的事物自然显露出来，而无须狂热强加各类意图。如此的设计方法能够产生从容的方案，促进场所空间和其使用者之间的对话。

这本《实践景观》设计作品专辑是大地景观事务所二十年来的经验汇集，展现出某些代表性项目的精髓，也表达着事务所对景观的研究、省思、看法以及事务所本身的特点。尽管对于将设计方案进行一般性的分类与定义，事务所向来持着保留态度，此书仍然试图通过四个不同领域来提出一种阅读景观方案的可能性：园林、历史、城市和自然。

Our Culture

Experience determines each of our interventions.

TERRITOIRES' projects are the result of a constant search for surprise and astonishment, designed with a caring consideration for the users of these spaces. This experiment with the unexpected falls within the agency's daily activities, as a laboratory analysing the land's contrasting scales, eager to search further afield to find its own answers. Understanding the complexity of a project is also a way of legitimising every detail, to make every aspect of the project reality-proof, and to find its place within the world.

TERRITOIRES claims a strong attachment to Culture: shaping landscapes and reflecting its population's intelligence and inventiveness. This attitude is reflected in a clear, simple yet precise aesthetic, where each design gesture captures the horizon, emphasizes contours, reveals the light, and unveils a story. As such, form is not the objective in our approach, but rather a tool to translate the site's requirements. This formal, measured and trustworthy approach, enables coherence and allows the land to express its peculiarity, revealing what is "already there" without attempting to impose. This inner conviction generates considerate projects, willing to produce a dialogue between places and their inhabitants.

"In Practice" is a collection of experiences focusing on TERRITOIRES' projects over the last twenty years. It offers a close analysis of the agency's large scope of research, reflection, responses and personality. Although it is normally reluctant to define its projects within categories, TERRITOIRES has presented its work under four possible headings: Gardens, History, Towns, and Nature.

gardens

花 园

一个花园代表着一个经验，一个属于生物世界的经验，展现出植物、动物与生态环境之间的繁复关系。

在本章介绍的项目中，这个糜集蠢动的生物世界被加以整理和赋予些许纪律，以便能够巧妙适宜地与周围环境相结合。大地景观事务所（TERRITOIRES）所设计的花园总是与所处的土地产生互动关系，而非一种抽象概念的呈现。这些方案充满着密度与惊奇，以简约的方式带领人们观赏生物世界。

A garden is an experience of the living world, of the complex relationships between animals, plants, and the environment they live in.

This biological swarming is disciplined and organised to cleverly blend in with the surrounding world. TERRITOIRES' gardens play with the territory without leading it into abstraction. These projects create density, surprise, and provide a simple perspective of the world.

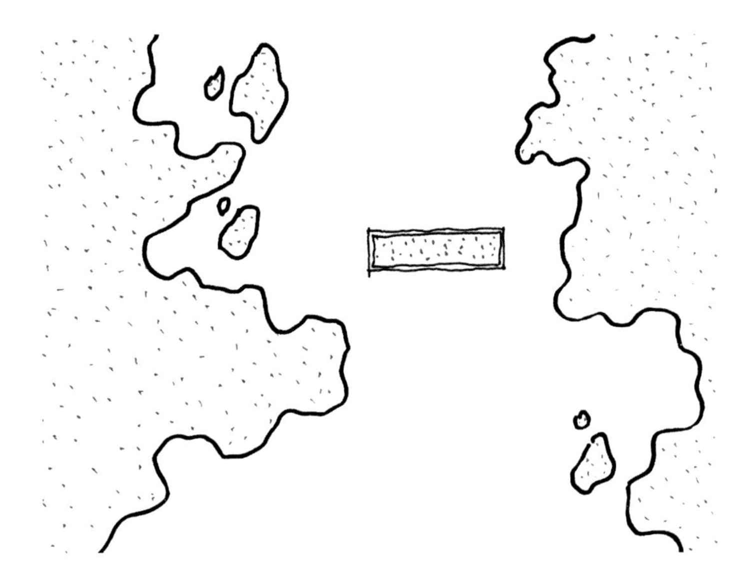

庄园性格
domaniality

法国 克莱尔方丹–昂–伊夫林 / 2016

Clairefontaine estate
克莱尔方丹庄园

位于大巴黎地区的克莱尔方丹庄园是一个充满神奇色彩的场所，在此缔造出法国足球的发展历史。此建筑群完全用于服务足球运动，一切设备皆为满足足球专业人士和爱好者的需要，其中包括一个国家培训中心。然而此地成名的原因是：它被当作是法国足球队的"家"。即将在法国举办的 Euro 2016 欧洲杯足球锦标赛，将会把所有人的目光吸引至此。

在略超过一世纪的时间里，此基地以混合的方式产生演变。伴随新的培训与会议中心的建造而进行的景观方案，唤醒了这个场所的基本特色，并且也让基地的庄园性格浮现出来。景观方案在回应多种使用功能与展示功能的同时，也为基地重新建立起整体性，借此彰显这个卓越运动园区的特性，并使之得以永恒持续。

项目的景观规划围绕着一个壮观的橡树林而发展，并且让人得以感受美丽的丘陵地势，带给整个基地一种既庄严又宁静的氛围，有利于高水平运动所需要的全神贯注。球员们的规律出现仿佛为此公园环境带来一道柔和的张力，强化出这个运动圣地的特殊性格。

The Clairefontaine Estate near Paris is a mythical place that witnessed the evolution of French football. The complex, which is equipped for amateurs and professionals alike, is entirely dedicated to the Beautiful Game. During Euro 2016, which will be hosted by France, all eyes will be on the national training centre, best known as the "home" of the France national football team.

In little more than a century, the site has evolved with mixed styles. The landscape design project, which is parallel with the construction of a new training and conference centre, highlights the fundamental qualities of the site and resurrects its identity as an estate. The landscape restores and preserves the concept and many uses of a campus with an athletic identity.

The estate, built around a majestic oak forest, offers stunning views of the hillsides. It infuses the whole site with a prestigious and quiet atmosphere which is conducive to concentration, necessary to the high-level activity that inhabits it. The presence of football players tilts the balance within the atmosphere of the park, highlighting the exceptional nature of this Mecca of sporting heritage.

上图：徜徉于谢夫勒斯上河谷自然公园景观中的运动园区
左图：方案研究

Above: The sports campus set in the Chevreuse High Valley Natural Park landscape
Opposite: Studies

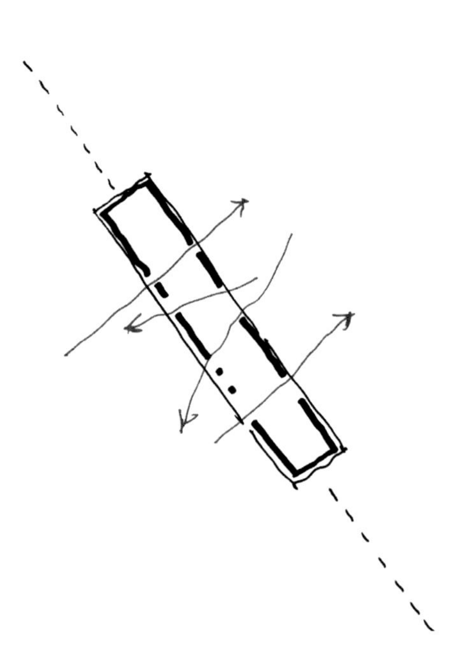

缝合
suture

法国 巴黎 / 2011

Serge Gainsbourg garden
塞尔日·甘斯布花园

塞尔日·甘斯布花园与周围环境建立了许多相当美妙融洽的关系。首先是与环城道（原为军事用途后来转为道路用途）的关系，花园正坐落于覆盖道路的平板上。此环城道是巴黎城市发展的历史见证，因此花园特地保留了个漏空区域，让人得以感受环城道的存在，延续城市的历史痕迹。一个观景平台将此每日300 000车流量的道路转化为一场场演出：一波又一波的机动车浪潮不断地填充又散去。

花园还扮演着为巴黎与郊区之间进行空间缝合的工作，甚至借助景观视野将关系推向更远的丘陵地。此花园成为行人穿梭和交流的场所，因此步行路线的规划成为城市与郊区联结结构上的重要元素。

“在城市建立自然空间”此概念的合理性已经逐渐显得普遍而平常，然而对巴黎里拉城门而言，此理念的实现却因为环城道路的存在而更具必要性。事实上，道路是生物多样性的极佳传播媒体，野生而四处旅游的植物种子借助车道气流的传送，自然地来到此花园驻留。这个方案邀请大自然在此自发生长，自行编织风景。

The Serge Gainsbourg Garden's essence owes much to its ability to forge meaningful relationships with its surroundings. At the beginning was the Périphérique, a ring road with military history, which has witnessed the historical development of Paris. The "Périf'" (ring road) is now canopied over by the garden and an esplanade, at the Porte des Lilas. The project preserves and maintains this urban outline through the void it sustains. As for the panorama, the daily motorized flow of 300,000 vehicles offers an impressive sight.

The garden provides a space for transition and exchanges, enabling the link between Paris, its suburbs and beyond. Pedestrian routes play a vital role in this interaction.

Finally, the presence of the ring road explains the choice of location at the Porte des Lilas, the legitimate yet banal idea of a natural space in the city. The road is indeed a formidable vehicle for biodiversity borrowed by winds carrying wild travelling seeds, naturally deposited to the garden. The project invites nature to spontaneously develop and grow its own landscape.

上图：具有活动用途的花园
右图：供人穿越的花园

Above: The inhabited garden
Opposite: The traversed garden

111.60
113.88
Avenue René Fonck
Périphérique extérieur
Périphérique intérieur
Bretelle de sortie
Périphérique intérieur

上图：观景台，一个观看巴黎广阔景观的场所
左图：花园坐落于覆盖着环城道的平板上，底下汽车川流不息

Above: The panoramic viewpoint, an observation point overlooking the vast Parisian landscape
Opposite: The flow of the ring road canopied by the garden

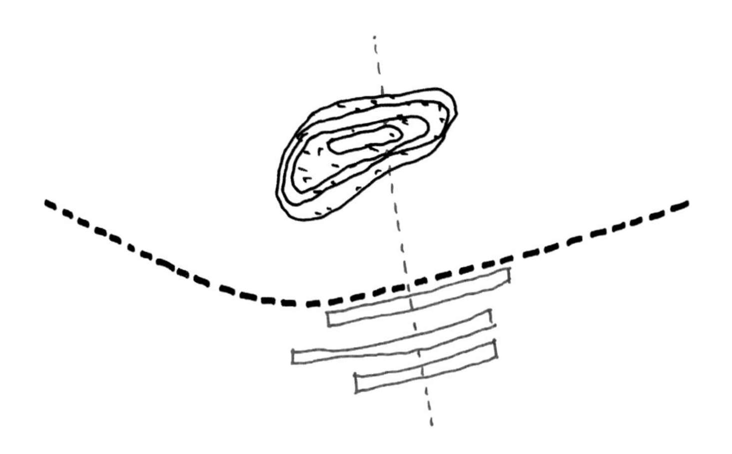

框景组构
composition

法国 莫罗 / 2011

Belfort-Montbéliard new TGV station
贝尔福-蒙贝利亚尔高速火车新车站

车站是功能性空间，然而贝尔福-蒙贝利亚尔的新车站面对连接孚日的蓝线火车却提出了一个十分独特的措施。火车世界的运动状态遇上了弗朗什-孔泰地区风景的稳定性，促成了一个崭新的方案，与大自然维持着紧密的关系。

位于森林边缘的车站基地具有独特地貌，享有面对兰花山丘、邻近与远处森林以及孚日山脉的广阔视野，如此卓越非凡的地貌在方案构思中占有决定性的地位。这些景观视野不仅受到保护，同时也被揭露和彰显出来。一些观景小栈沿着参观路线而设置，提供游客静心观赏美景的时刻。1200个停车位的规划则着重于将地理特征展现出来，同时也将人们深置于弗朗什-孔泰地区的大自然中心。

地面径流的管理设施遵循着景观的自然状态，通过一系列斜沟呈现出从最干旱到最潮湿的连续生态环境。最后一个斜沟对准谷底线而设置，成为提供车站旅客欣赏的景观花园，也是沟中动植物的生态花园。这个原本十分技术性的项目被转化成为人们理解大地景观的工具。

Train stations are functional. Yet the new Belfort-Montbéliard station is located in a very unique spot, facing the blue line of the Vosges. The meeting of a world of motion and a countryside qualified mostly by its stillness defines a project with a relationship to nature at its very core.

The landscape design project aims to protect, reveal and intensify its remarkable panoramas: the Butte aux Orchidées, the surrounding forest and the Vosges mountains in the distance. Many panoramic viewpoints are dotted around the travellers' itinerary across the 1200 space parking area, showcasing the surrounding Franche-Comté landscape, placing man at the heart of nature.

The rain water management device, consisting of a succession of dry to wet reservoirs, offers a range of very natural looking water ponds and an overall natural environment. One of them provides a garden area for station users, and is home to the local flora and fauna, embedded on the valley thalweg. Its extraordinary attributes have enabled this highly technical project to define a new understanding of the area.

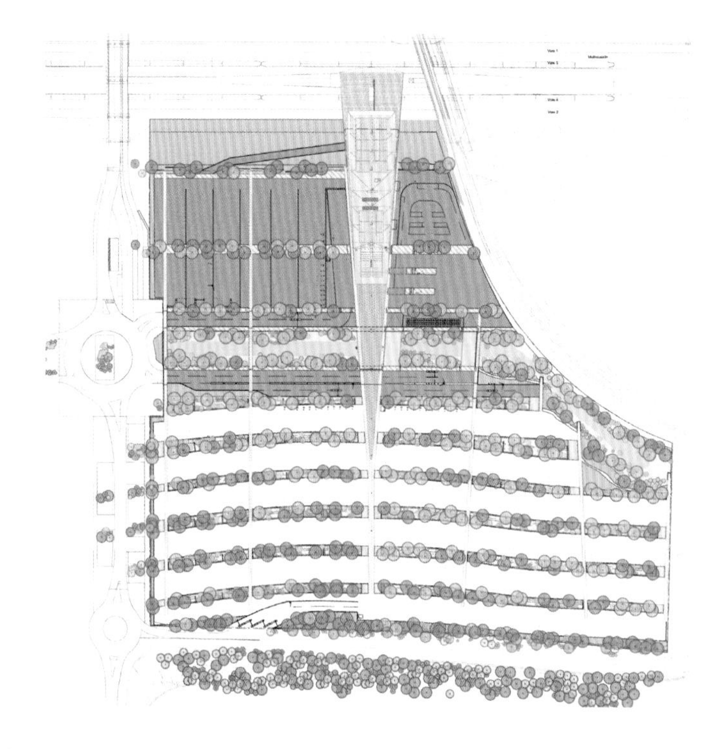

已存在的地形延伸于停车场的配置构图上

The car park design extends the existing topography

6E

步行路径上的设施将此地特殊的地理形态凸显出来

The presence of the path reveals a singular geographical location

绿洲
oasis

西班牙 马德里 / 2009

Valdebebas park
瓦德贝巴公园

在位于马德里北边、巴拉哈斯机场附近的新街区发展当中，瓦德贝巴公园成为第一阶段建设的关键空间。此一新街区具有提供国际中转相关服务的潜在能力，因此其发展融合了住宅与商业的双重目标。这个位于贫瘠丘陵地上的辽阔公园不仅必须转变为一片自然林地，同时也必须提供马德里居民一个休闲娱乐公园。

通过设计竞赛，大地景观事务所（TERRITOIRES）的方案提出了一个绿洲化的过程，建议建造一系列的蓄水墙来将雨水持住，而后慢慢释放。这些设置在建筑物脚下的灌溉机制既温和且规律，得以逐渐产生绿洲园地。方案刻意安排以不同高度的植物来塑造景观，同时也使较高的植物能够过滤阳光，保护较为矮小脆弱的植物。

这些阴凉的绿洲为人们提供娱乐休憩的场所，并且使得公园里的建筑物与某些活动得以集中：餐厅、亭式小卖部、表演空间、分享式花园……。这些绿洲同时也通向一些观景台，让人们得以眺望远景以及公园的其他区域：种植着橄榄树、野草莓树、岩蔷薇、爱神木……

The Valdebebas Park project is the first step in developing a neighbourhood to the north of Madrid, near Barajas airport. Due to its high potential for international transit, this future neighbourhood will require both a residential and commercial use. Its large park, inclined on a vast barren hill, will provide a natural wooded area and recreational park for the city of Madrid.

TERRITOIRES came up with an oasification process for the competition proposal, creating reservoir walls to slowly release trapped rainwater. Oases are thereby formed by the gentle and regular irrigation of gardens below the buildings. The landscape is planned into different layers of vegetation to filter the sunlight and protect the most vulnerable plants.

These cool shaded areas host the garden's events and leisure facilities, including restaurants, kiosks, concert spaces, community gardens, etc., which are open to the public for entertainment and relaxation. These areas are surrounded by panoramic viewpoints of the stunning park and landscape, including olive trees, scrub arbutus, cistus and myrtle...

绿洲园地将公园里的活动场所聚集在一起
第 31 页：蓄水墙的建造使绿洲园地得以逐渐产生

The oases provide a focus for the park's leisure areas
Page 31: The construction of reservoir walls enables an oasification process

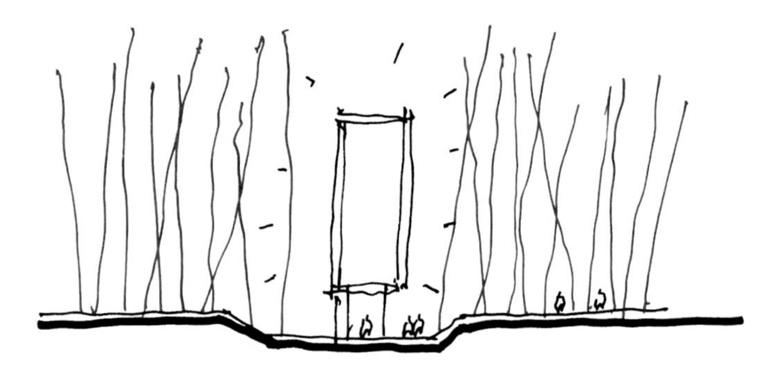

沉积
sedimentation

法国 葛雷 / 2003

Tilleuls square
椴树广场

城市空间不断产生沉积作用，旧的痕迹消逝，新的接踵而来，为城市带来复杂的地面状态。椴树广场原本是为了取代一个旧花园而成为散步道，而花园本身则建造于中世纪的护城壕沟上。

对这个沉积过程的观察，导致方案决定让历史重现，让城市的记忆得到保留。这个设计意图特别采用了黑色材质来体现，明显有别于颜色较为浅淡的当地石灰岩，以避免产生任何混淆的可能性。整体空间组织遵循着旧花园和中世纪壕沟的明确遗迹，对于未经考古学家证实的形迹则予以排除。

由此产生了一个具有当代感的混合方案，同时展现着现代特质与历史情感。方案还开发了一些散步路线或休息地点，并且将一个已消失的旧亭子象征性地重新建立，也在广场与葛雷老城之间进行一些细致的整治。

本方案荣获2004年城市空间整治奖。

All urban spaces are sedimented. Traces of the city fade, giving way to others, thereby building the complexity of its soil. The Tilleuls square, a promenade built on a garden, was created on the old moat that protected the medieval town.

Research into this sedimentation drove the choice to retain the urban memory and allow history to transpire. This objective is reflected in the use of dark materials, which can be clearly distinguished from the lighter local limestone. The overall composition is based on tangible traces of the old garden and the position of medieval ditches, defined through the approval of archaeologists.

A hybrid contemporary project was thereby developed, combining its modernity and its attachment to history. The project redefines routes, intervals, a forgotten kiosk, with meticulous attention to detail, between the edge of the square and the old town of Gray.

Prize for Urban Planning 2004

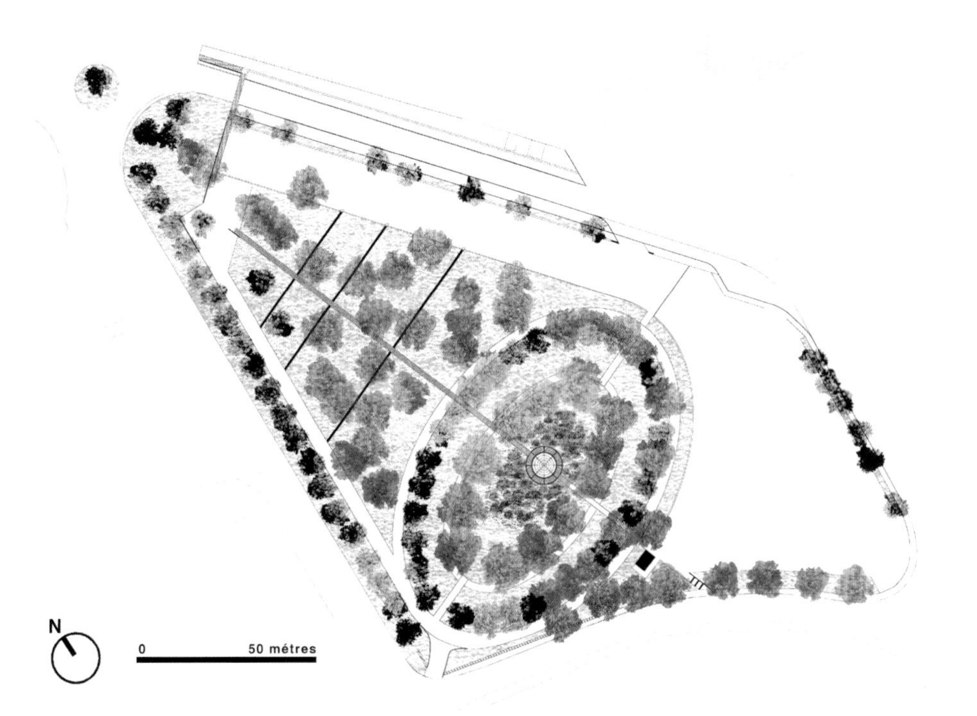
N
0
50 métres

上图：灯笼凉亭以象征性的手法展现历史的延续

Above: The lamp-kiosk symbolically concludes a gesture of historical continuation

histories

历 史

在大地景观事务所（TERRITOIRES）经手的环境项目中，文化遗产的课题犹如丰厚的土壤，带领方案完全融入一个场所和一段历史的现实之中。

历史是方案构思的前提，却非方案设计的目标。因此历史研究的工作丝毫不带怀旧情感，却充满着建立一种连续性的渴望，期望延续某种永恒的轨迹。在此种历史态度下所实现的方案，有时具有相当的挑动性格，勇于对时间观念提出质疑。

Legacy transpires in TERRITOIRES' projects like a framework, providing the foundations for each project, which is strongly rooted in the existence of a site and its history.

History is a prerequisite but not a finality. The work shifts away from nostalgia and defines itself by its desire to express a certain continuity. This relationship with history can bring about inspirational projects, questioning the notion of time.

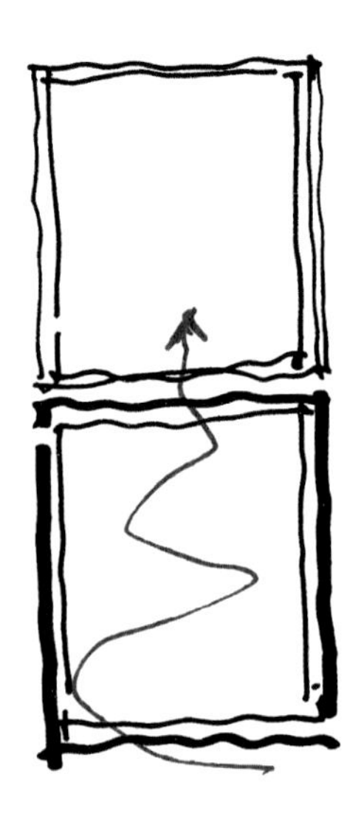

记忆
memories

法国 莫城 / 2011

museum of the great war
一 战 博 物 馆

在1914年9月马恩河战役时，莫城高原上演着残暴对峙的景象，此后则剩下寥寥无几的历史痕迹，其集体记忆如今也颇受折损。2011年，在美国纪念碑脚下崭新设立的一战博物馆悬突于丘陵斜坡上，在战役的地理环境中重新还给历史一个位置。博物馆的场景设计以时间和记忆为轴，形成一个环状路线，引发造访者的历史情感和全神贯注的精神，塑造出极为特殊的氛围。

博物馆公园由一个停车场和广大花园所组成，其景观规划延续着室内场景设计的精神，让人们与户外场所产生对话。造访者被引导穿越一片雪松林，犹如步兵一般前进，不禁对于历史以及创造历史的人们感到肃然起敬。户外景观仿佛是造访博物馆之前或之后的心理调节过渡空间。

走出博物馆之后，在花园中的漫步成为另一种转化过程，从室内的庄重氛围来到一个承载着过往训示的当下时刻。

In September 1914, the plains around the city of Meaux were the scene of violent fights during the Battle of the Marne. Now, only a few traces remain alongside a somewhat faded collective memory of these events. In 2011, on the hillside close to the American monument, the new Museum of the Great War restores this history to its geographical context. Its museography draws a path, creating a loop in time and memory. The museum sets an atmosphere by evoking the emotional charge that captures the visitors' attention.

On the outside, a parking and large park comprise the museum's grounds. The landscape continues the museum tour, setting the scene for an exchange between visitors and the environment. In a spirit of simplicity and without wishing to recreate nor mimic the past, visitors are led through a cedar wood following the way the infantry would have moved, allowing them to understand and gain a new found respect for those who took part in the historic events. The landscape acts as a psychological conditioning chamber before and after the visit.

Upon exiting the museum, the park's promenade creates another transition, leaving solemnity behind, leading to a present enriched with lessons from the past.

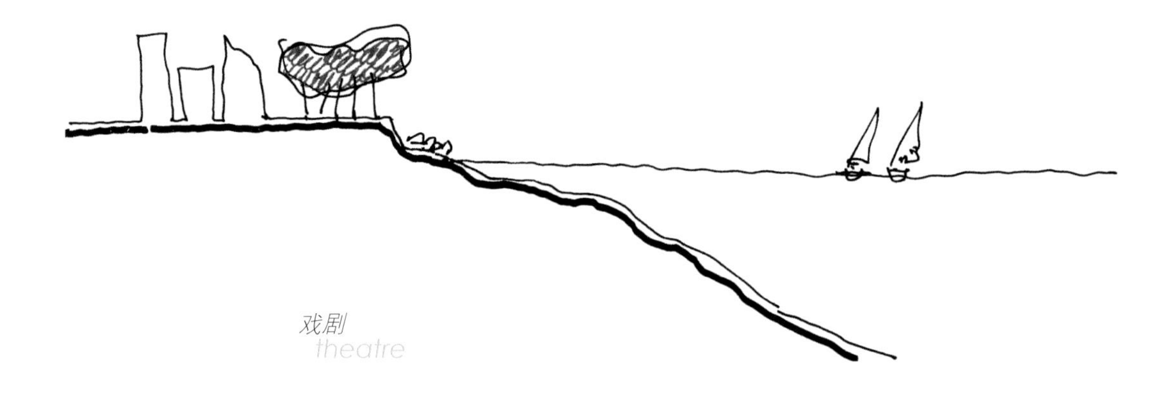

法国 尼斯 / 2010

promenade des anglais
英国人散步大道

尼斯是蔚蓝海岸上一个神话般的城市，由于地处地中海和山区腹地之间，棕榈与橡树便同时存在着。这个位于边界地区的城市，自然而然地受到前来度假的众多外国人的影响，其“天使湾”和著名的英国人散步大道成为一个令人觊觎、充满异国情调的场所，犹如一座排满戏码的露天剧场，熙熙攘攘的人群，有的前来观赏，有的前来展现。

著名的地中海建筑师鲁迪·里乔蒂与大地景观事务所（TERRITOIRES）携手合作的方案旨在将城市建造的特征与地理环境的特色显现出来。英国人散步大道的天然美景与人为活动固然已经存在，然而方案仍然提出崭新的设计，来将已随着时间而消逝的高贵庄严的特性重新找回。这条散步大道是面向海洋与城市的眺望台，同时犹如一座提供所有人参与演出的舞台。

景观设计所呈现的场景由一系列不同的空间、惊奇、情节以及丰富创新的张力所组成。植栽计划为海湾塑造出全新的环境，使散步道在光影效果中重新苏醒，并且为城市与海洋提供了多样的框景。人们的目光因为这个新的规划而产生转变，海湾本身也因而成为瞩目的焦点。

Nice is a mythical city on the Côte d'Azur. Between the Mediterranean and mountainous hinterland, palms and oaks live side by side. As a border town, Nice has always been sensitive to the contributions of foreign visitors. This coveted "Bay of Angels" exudes exoticism, and together with its famous Promenade des Anglais, can be compared to an open air theatre, whose permanent stage show is a draw for all to watch or shine.

The project, led by Mediterranean architect Rudy Ricciotti and TERRITOIRES, reveals the true spirit of the place and the singularity of its location. Its natural beauty and drama have already paved the way, but a new design aspires to restore the somewhat faded elegance and dignity of the Promenade des Anglais. The Promenade is transformed into a balcony overlooking the sea and the city, a theatre where all actors make full use of the stage.

The landscape design becomes scenography, a series of sequences, surprises, intrigues, where the balance is in constant flux. Flower beds introduce a new framework to the bay. The Promenade comes alive with the interaction of shadows and light, creating extensive compositions around the sea and the city, and projecting the bay to centre-stage.

地面和树木被当成装饰元素来处理，以重新赋予这条散步道已被遗忘的阔气排场

The ground and trees provide decorative elements that restore the Promenade to its former splendour

GRESCO

游戏
play

法国 苏镇 / 2006

town hall gardens
镇 公 所 花 园

这个镇公所花园的设计展现出游戏的精神。与历史玩场游戏（法国第二帝国时期典型的专区政府，1843年由建筑师克劳德·内桑所设计建造，他同时也构思了巴黎第16区的区政府建筑），并且与某些学院派做法保持一定距离，这是经过刻意选择的景观设计方向，以提出一个突出而清晰的方案。

幽默与游戏趣味将方案导向一个十分严格的空间组织工作，以便在这些空间中塑造一些异常状况，突破了此类建筑惯于搭配的花园意象。向着海绵状矮墙倾斜展开的大草地上种植着植物，此草地能够汇集雨水，并且灌溉墙上的植物以及下方水池中的植物。一系列平台与墙面的设置不仅产生视觉趣味，也将一些在巴黎地区极为少见的地形特色彰显出来。

由伊夫·莫雷尔设计的马形雕塑让人联想起古时专区政府的马厩，同时也在如此一个庄严的场所为孩童们创造出令人惊喜的游戏空间。

Here, experimentation and play go hand in hand. This bold and structured project embraces the site's history (the emblematic old French Second Empire subprefecture, built in 1843 by the architect Claude Naissant, who also designed the Town Hall of the 14th arrondissement in Paris) and somewhat distances itself from the more formal and conventional rules and traditions.

In the spirit of wit and playfulness, the project's detailed composition has produced some unexpected dynamics. The garden is detached from the traditional imagery associated with this type of architecture. A lawn planted towards a cavernous wall collects runoff and feeds the plants on the wall and in the basin below. The creation of terraces and walls provides a range of views, emphasising its singular location in the Parisian region.

Equestrian statues by designer Yves Morel reflect the presence of the former subprefecture's stables, offering children an unexpected playground in this unusually solemn place.

本方案结合了两种花园空间：具有硬质铺面和骑士精神的红花园（旧宪兵队营房），以及设有繁茂植物和水生环境的绿花园（旧专区政府）

This project draws two areas together: the red mineral garden, inspired by an equestrian theme (a former police station), and the more verdant and aquatic green garden (a former sub-prefecture)

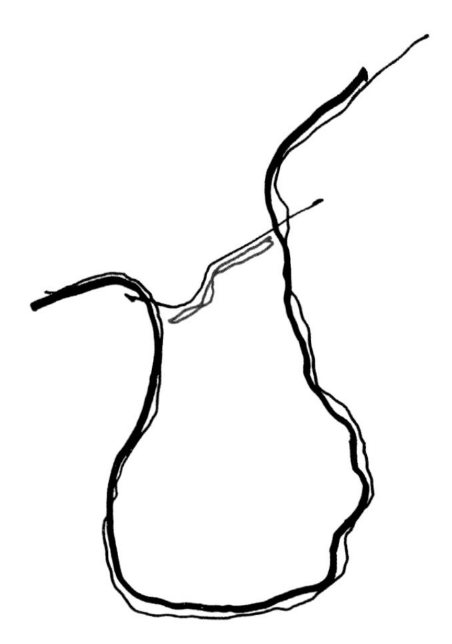

传承
heritage

58

法国 西–苏尔–索恩–圣阿尔宾 / 2006

Saint-Albin canal
圣阿尔宾运河

圣阿尔宾运河是19世纪时由工程师菲利普·拉科代尔所开凿的一条航用渠道，总长681米。这条水利工程的建设目标在于让索恩河的水渠能够在1870年发生的德法冲突之后提供航行功能。当时德国对法国关闭了前往莱因运河的通道，并且使其断绝了与欧洲北部的联系。如今此运河作为游艇的航行使用。

此水利工程具有超卓品质，其稳固的圬工砌造、顺着地形地势而产生的曲线与反曲线造型，使其得以被列入法国历史文物保护的注册名单。这个出自同一人之手的建造系统，使运河能够呈现出合理统一的建筑、工程结构体、土方和装饰。

为了彰显运河而作的研究，引导着方案设计对已存的景观模式进行简约而忠实的延续与更新。事务所在重显运河的过程中查询了诸多史料文献，对历史痕迹赋予高度的关注，并以谦虚的态度和清晰易读的方式重新展现这个场所的固有结构。

The Saint-Albin Canal is a river tunnel dug in the 19th century by the engineer Philippe Lacordaire. The purpose of this 681 metres long work of art was to turn the Historical Canal on the Saône into a waterway, after the 1870 conflict, when the German authorities closed access to the Rhine Canal, and with it, the gateway to Northern Europe. Today, it is used for recreational boating.

The quality of workmanship, which earned its recognition as an historic monument, lies in the impressive stonework, curves and counter-curves across the relief, marking its territory. Architecture, art, landscaping and ornaments come together within one design.

The study of enhancement is the true and simple renewal of a landscape model. The agency's restoration work was guided by the analysis of recovered archives. This requires great attention to fragments left by heritage and a certain humbleness, as designers, to enable its former structure to become visible once more.

法国 蒙梭–雷–敏讷 / 2006

Chavannes's coal washing plant
沙 畹 洗 煤 厂

这个建造于1923年的沙畹洗煤厂位于盛产煤矿的地区，产业衰落之后则变成一片巨大的废墟，如今这个工业圣地已经被列入法国历史文物保护的注册名单。MVRDV建筑与城市事务所和大地景观事务所（TERRITOIRES）联手赢得了沙畹洗煤厂改造的设计竞赛，此竞赛旨在研究基地转型的可行性，在保存城市记忆和贯彻生态发展的双轴线下，提出各种可能的假设方案。

获选方案提出了一种与众不同、具有现代性与可持续性并且运用较少人为手段的方法，来将这个伟大矿产时代的历史融入地区的当代环境之中，使洗煤厂不再成为孤寂的幽灵，也使区域的发展能够围绕着它而重新起飞。这不仅仅涉及让大自然重新进驻废墟，并且必须把植物化的过程纳入控制之中。计划范围仅限于曾经遭到污染和已经去污染的地块。

老旧的洗煤厂期望拥有现代废墟的面貌，不仅能够骄傲地展现城市的历史根基，也替因为主要产业没落而经济受创的整个地区带来更新的动力。设计方案参考了目前欧洲的许多大型基地的转型经验，不局限于地域性的养分与资源。

The Chavannes's Coal Washing Plant was built in 1923 in a coalfield area. This industrial cathedral, which presently stands as a vast wasteland, has been listed as an Historic Monument. The competition, which was won by MVRDV agency in association with TERRITOIRES, aims to study the feasibility of a site conversion, considering several options for the preservation of urban footprints and integrating the notion of eco-development.

Finally, the project promotes a different, modern, sustainable, perhaps less artificial way to include this relic of the region's great coal mining era in a contemporary setting that would allow the ruin to move away from its haunting presence. It is not simply a matter of letting nature reclaim a wasteland, but to control its revegetation. The scope of development is limited to the containment and decontamination of polluted site.

The old coal washing plant will achieve the appearance of a modern ruin. It displays the powerful historical foundations of a city while embracing the renewal of an entire region affected by the decline of its founding industry. Through this, the redevelopment is consistent with major European programmes, reaching far beyond local boundaries.

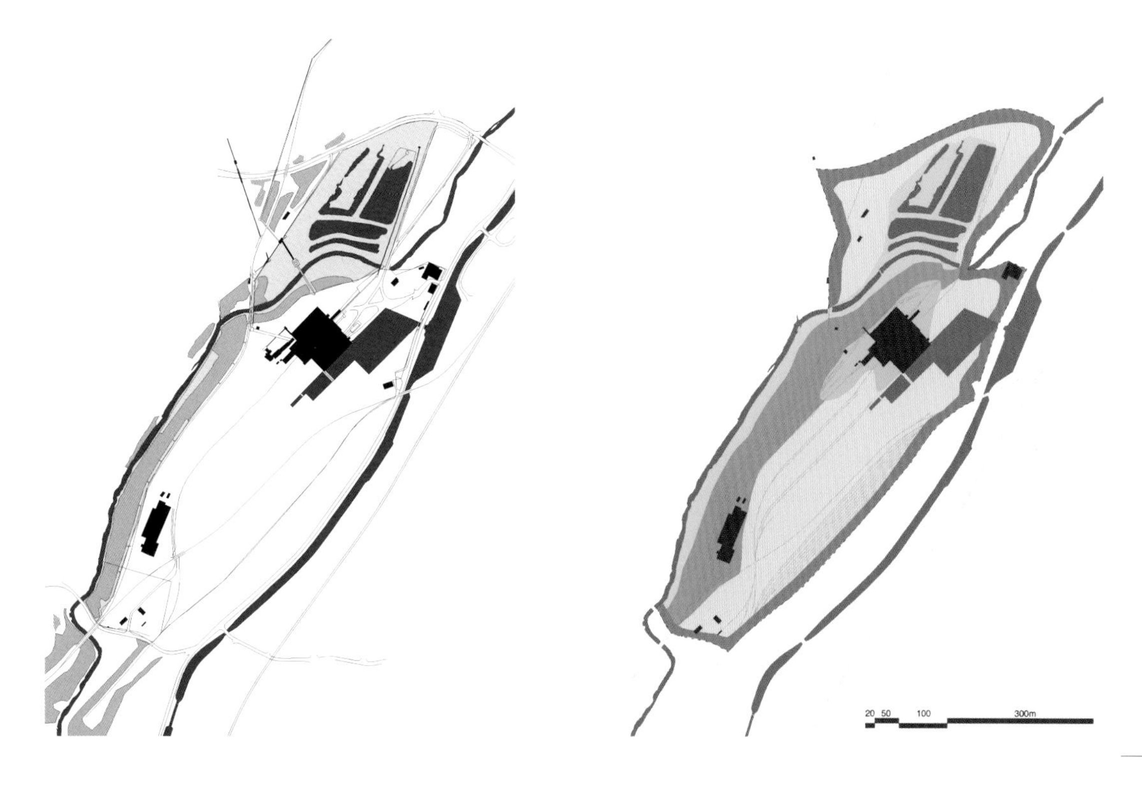
20 50 100 300m

上图：工业废墟（1）与热带丛林的抽象概念（2）融合，从中浮现出“矿产工业殿堂”（3），以达到大自然重新收复废墟的结果（4）
右图：植物重新收复失土的过程

Above: The industrial wasteland (1) incorporates the abstract concept of the jungle (2), with its emerging "temples of the mining industry" (3) letting nature reclaim the ruins (4)
Opposite: The vegetation reclaiming process

20m
15
10
4
2
5
10
20 ans

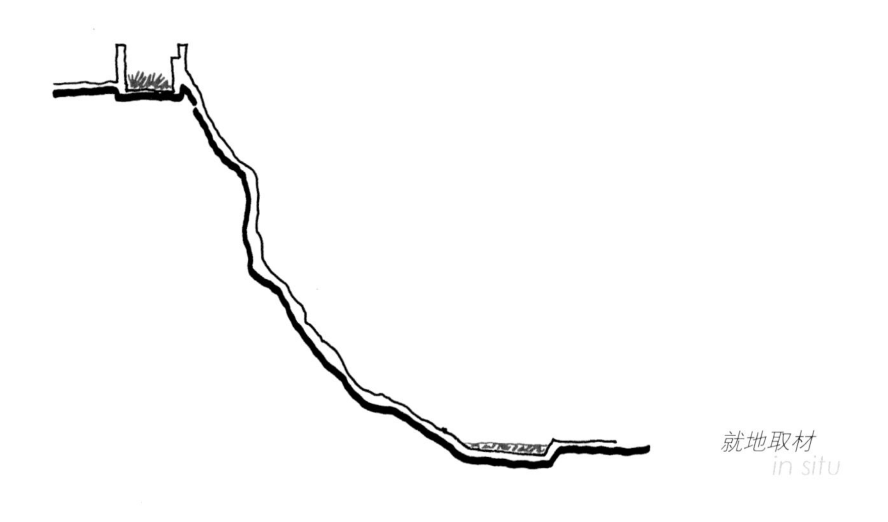

就地取材
in situ

法国 贝桑松 / 1995

Vauban citadel's aquarium
沃 邦 城 堡 水 族 馆

贝桑松城堡起源于凯尔特人所创造的漫长历史，并且于17世纪在沃邦王朝的辉煌建筑成就下奠定了如今的形貌。城堡的600米城墙与地形巧妙结合，不仅居高眺望城市，也拥有壮丽独特的景观，整个宏伟的基地被国际教科文组织列为世界遗产。城堡内设置了不同主题的博物馆，其中包括了一个专门为杜河沿岸自然湿地而成立的水族馆。在高于城市与河流100米之上的城堡干燥环境内建设一个湿地花园，意味着必须动用许多人为设施才能塑造出水岸的天然环境。

此地所有的参观流线都高于地面而设置，木板通道延伸了博物馆立面的木造装修，将项目的潮湿特质更为彰显出来，并且在这些木质“线条”之间也设置了各种不同的生态系统。为了尊重这个保护建筑的原貌，此方案工程的实施运用了“可逆向”原则以及使用元件的“可读性”原则。

在此方案中，只有一个水塘被重新组构以提供水生植物与动物生长环境，其他的设置则不刻意模仿自然环境，而是直接使用在基地可以就地取材的环境。方案选择了以相对密集的方式来重新组构一个丰富多样而且能够适应城堡建筑的生态环境。

The Citadel of Besançon has a long historical background, initiated by the Celts and glorified by the architectural genius that was Vauban in the 17th century. The 600 metres of ramparts follow a relief overlooking the city and its unique landscape. This monumental complex is a UNESCO World Heritage Site. The fortress houses several thematic museums including an aquarium dedicated to the natural wetlands of the river Doubs. The creation of a wet garden in this dry, fortified, 100 metre high exceptional site above the city and river requires a few clever tricks to reproduce the natural atmosphere of the banks.

All movement is thereby lifted from the ground. The wood decking amplifies the project's wet nature. It extends the woodwork design on the museum façade, allowing different ecosystems to combine within these "bands of wood". This project adopts the principle of reversibility within the renovation, creating a transparency which conserves the integrity of the monument.

One single pond has been restored to host a range of aquatic flora and fauna. Rather than attempting to imitate a natural environment, its surrounding features aim to reconstruct a thicker natural abundance, adapted to the Citadel's architectural context.

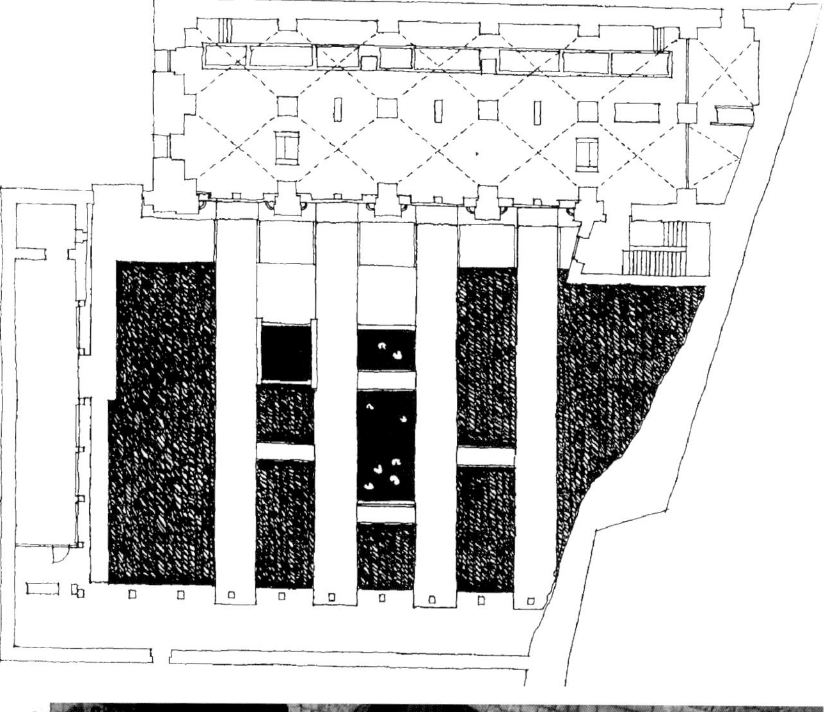

上图：杜河景观在此以线性序列的方式重现
右图：沃邦城堡建筑为花园提供了独特的环境与视野

Above: The Doubs landscape featuring linear sequences
Opposite: Vauban's fortifications form a part of the garden's setting and horizon

cities

城市

有些城市所处的地理环境相当复杂，在城市与大地之间，彼此关系不断受到强化，其张力在极度紧绷后松弛下来，以至于景观在人们眼中成为失去活力的载体。大地景观事务所（TERRITOIRES）设计的城市空间，旨在成为一种交流的场所，与周围的环境或经过其中的使用者产生对话。

这些城市空间不断自我创新，充满实验精神和愉悦氛围，景观便也自然地入驻其中。

Cities are sometimes located in complex geographies. The relationship between city and country has developed and distended allowing that landscape to become a static backbone. In contrast to this, the cities designed by TERRITOIRES invite and exchange with their surrounding and passing environments.

They are inventive and in perpetual growth, becoming spaces for experimentation and pleasure, which draw in the surrounding landscape.

瑞士 梅林 / 2018

Les Vergers eco-neighbourhood
果 园 生 态 小 区

位于梅林镇的果园生态小区处于特殊景观环境中，使得方案得以借此建立一个具有创意的城市街区。由于基地面对着汝拉山脉，方案特别着重小区与邻近乡野的关系，以既不怀旧也不矫饰的手法，为小区塑造出独特的城市环境，不仅使得农业在其中占有一席之地，也借此赋予空间特色，形成一个田园小镇。方案赋予园艺与城市农耕的地位、对空间管理的看重，使得居民们也卷入了一场非凡的经验：亲自参与他们生活空间的建设。

居民参与是方案的关键要素，是社会学家和大地景观事务所（TERRITOIRES）在项目前期便加以思考的重点。这个与居民共享的规划过程，让计划内容显得更为丰富，并且塑造出一个具有关照能力的“聪明”小区，也强化了居民的归属感以及他们与大地的关系。

在此项目中，可持续发展的观念并非作为装点之用，而是创造舒适城镇气候的方法。政治人物、居民、艺术家、管理者、建筑师、耕作者、园匠、商人等，共同在此创造出丰富而多样化的城镇生活，并且与瑞士的大地景观建立起美妙的关系。孩童们也在此同时受到城市文明与大地景观的熏陶。这个城镇小区不仅具有发展上的可持续性，同时也具有专注关照的性格。

The Vergers eco-neighbourhood invents an area within the city which is responsive to the landscape it inhabits. The project, facing the Jura Mountains, constantly strives to embrace its nearby countryside. Without nostalgia or sentimentality, it encourages the presence of agriculture within urban conditions, thereby defining the quality of the space. It is a rural town. Gardening, urban agriculture, and careful space management, engage its inhabitants in a unique adventure, taking part in the construction of their living space.

An upstream study conducted by sociologists and TERRITOIRES indicated that the inhabitants' involvement was crucial to the project. This shared work enriches the quality of the programme and enables the insightful creation of a city, adapted to its inhabitants' motives, thereby generating further connections with the land.

The concept of sustainable development is not intended as a veneer, but rather as an approach to creating a favourable urban climate. Policies, inhabitants, artists, managers, architects, farmers, gardeners, and retailers reinvent a rich and varied urban life, skilfully suited to the Swiss landscape. Here, children learn about urbanity and landscapes. The city rests in its permanence, and the awareness of its surroundings.

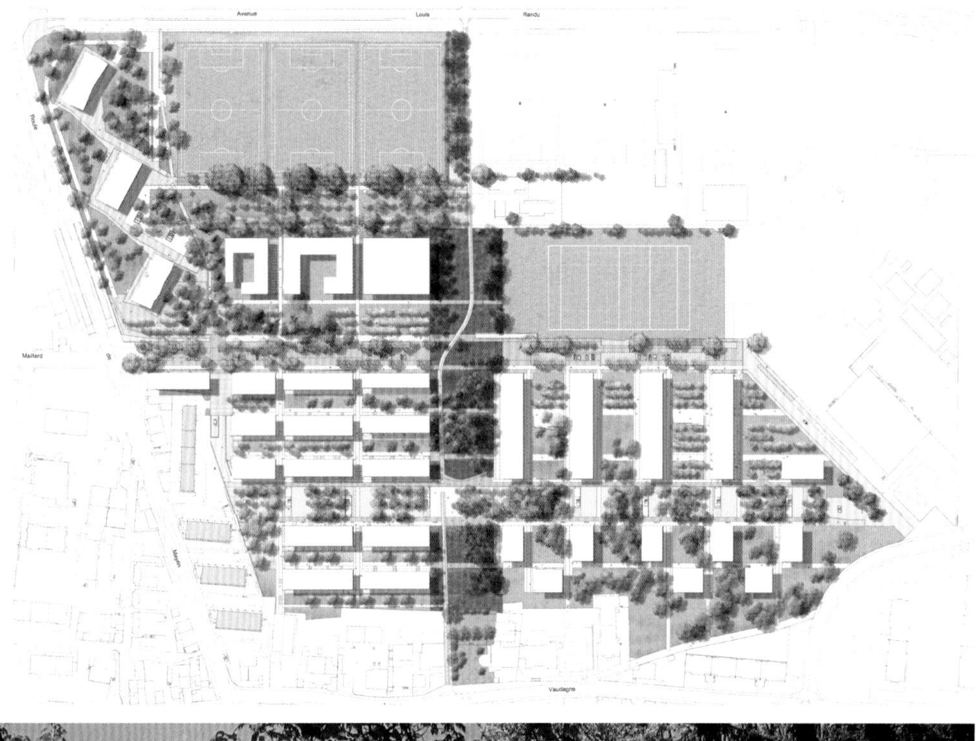
Avenue
Louis
Rendu
Route
Maillard
de
Meyrin
Vaudagne

MEYRIN
MEYRIN
MEYRIN
MEYRIN

一个为了保存乡村的尺度和作息活动而设计的生态小区

An eco-neighbourhood designed to preserve the land's physical dimension and practices

地质
geology

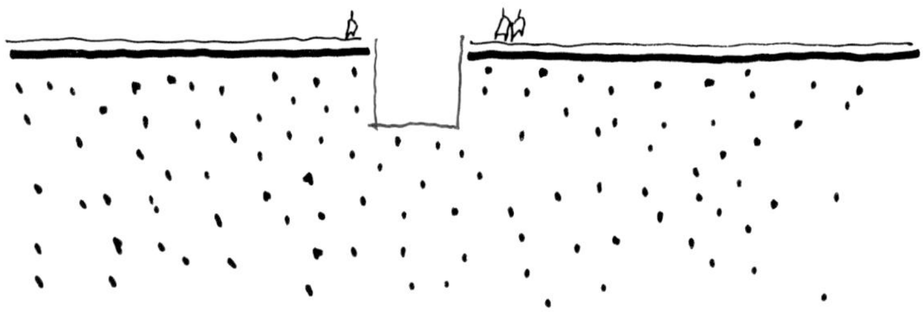

法国 米卢斯 / 2014

Neppert gardens
内贝尔花园

这个位于米卢斯的列斐伏尔旧军营改造项目，为沃邦-内贝尔街区带来深度转变的机会。此街区长久以来被视为是人迹鲜少的场所，其改造项目显得雄心勃勃：不仅要达到军事建筑遗产保护的目标，同时也涉及城市更新、一百多户住宅的建造以及公共空间和花园的开创。

城市规划师尼古拉·米其林（ANMA事务所）所提出的城市建设原则必须经历许多年的发展，并且循序渐进地将街区转变为名副其实的生活空间。以生态花园组构一个大公园的想法成为此景观方案的基础。广大的绿化公共空间不仅为居民提供了休闲活动场所，也成为连接该街区和其他城市区域的步行通道。

米卢斯城市建于莱茵河的旧河床之上。内贝尔花园位于建筑群的中心，以一系列主题花园的形式呈现，它们同时也形成一个能够将地面径流导向地底下的大型机制。花园里的承雨池调解了地表和地下的关系，并且显现出阿尔萨斯平原的地质特色。花园的植物也同样表达出莱茵河畔的土壤特性。

The conversion of the former Lefebvre military barracks in Mulhouse is an opportunity to fundamentally transform the entire Vauban-Neppert neighbourhood. The district, which was long considered uninviting, is at the centre of an ambitious redevelopment: the preservation of the military architectural heritage and urban renewal, the construction of hundreds of homes, public spaces and gardens.

Town planner Nicolas Michelin (ANMA Agency) has selected planning principles which will develop, transform and bring life to the neighbourhood progressively over several years. The concept of an inhabited garden park is fundamental to this project. Extensive planted public areas are available to residents and provide pedestrian links connecting the area to the rest of the city.

The Neppert Gardens are located within the heart of the residential buildings. This series of themed gardens is conceived as a comprehensive system capable of conducting water runoff into the sub-soil. Mulhouse is built on the former bed of the Rhine. The large volumes of accumulated pebbles lead water into the sub-soil. In the gardens the impluviums provide a channel between the surface and sub-soil, revealing the geology of the Alsace plain. The gardens' vegetation reflects the nature of this exceptional soil, unique to the banks of the Rhine.

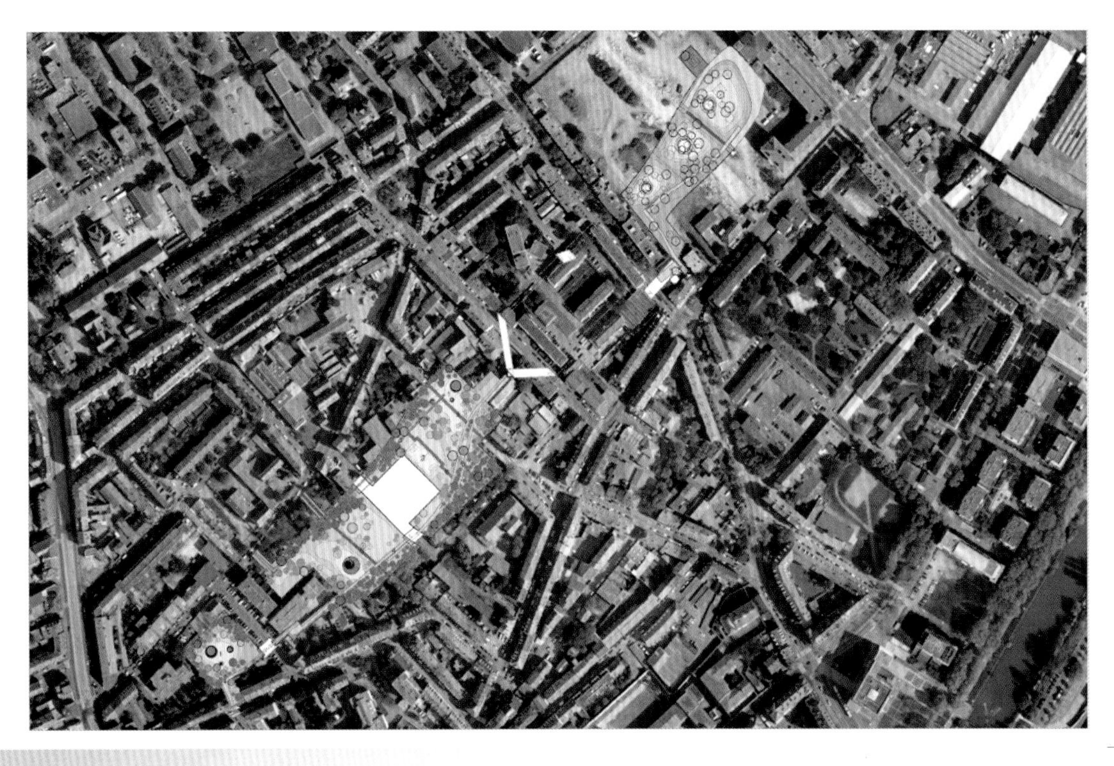

花园与承雨池将人们的目光导向被重新显现的土地

The garden and impluviums create a focus on unearthed areas

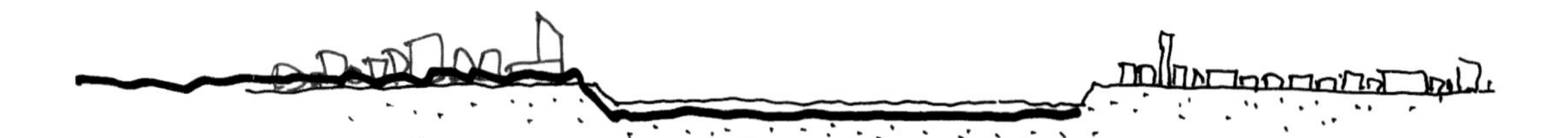

法国 波尔多 / 2012

bastide Niel
尼尔城堡

对波尔多城市而言，使河流两岸的街区产生连结一直是重要的城市整治目标之一。其河右岸长期以来遭受河水泛滥之苦，仅做为农耕、军事和工业等用地，成为城市活动的外围区域。这个河岸的开发，对波尔多城市尺度的扩大而言是至关重要的挑战，同时它也将赋予波尔多一个崭新的身份，成为名副其实的加龙河城市。

荷兰MVRDV事务所的建筑师们与大地景观事务所（TERRITOIRES）合作为此进行的前期城市研究，旨在为尼尔城堡建立一个指导计划。这是一个复杂的规划项目，牵涉着强大的政治、社会与经济推动力。此计划不仅致力于历史记忆的保存，同时也必须为铁路与军事废墟进行改造，使其成为一个真正的功能混合区域，建立一个理性发展和可持续发展的街区。

生态小区在流动性与能源的规划上必须具有雄心远见，在既有的城市基础上发展，以亲密型城市或高密度城市为蓝图，犹如古代城市的当代演绎。这个计划建议缩小道路宽度、设置柔性交通、交谊场所以及硬质铺面的公共空间，并且提倡能够保证日照的体量集中式建筑、一系列的主题花园和一些混合使用与集体使用的“口袋空间”（小规模开放空间）。

One of the main concepts for this project is to reunite its banks. For a long time, Bordeaux's right bank was subject to the whims of the river, and solely used for agriculture, military activity and industry, away from the city. Taking over the right bank is a shift in scale, creating a new identity for Bordeaux: a city on the Garonne.

The pre-operational urban study led by Dutch architects MVRDV in association with Territoires, established a strong basis for the Bastide Niel master plan. It is a complex operation, with strong political, economic and social motivations. The challenge is to preserve the past while transforming rail and military wastelands, ensuring an effective range of uses, building a controlled and sustainable area development.

The eco-neighbourhood needs to be ambitious in its mobility and energy goals. It begins with the existing traces that are translated into the concept of a dense but intimate city (a contemporary reinterpretation of an ancient city). This involves reducing road widths, introducing soft mobility and meeting areas, mineral public spaces, a volumetric architecture providing sunshine and a series of thematic micro-gardens, "pocket areas", intended for mixed and collective uses.

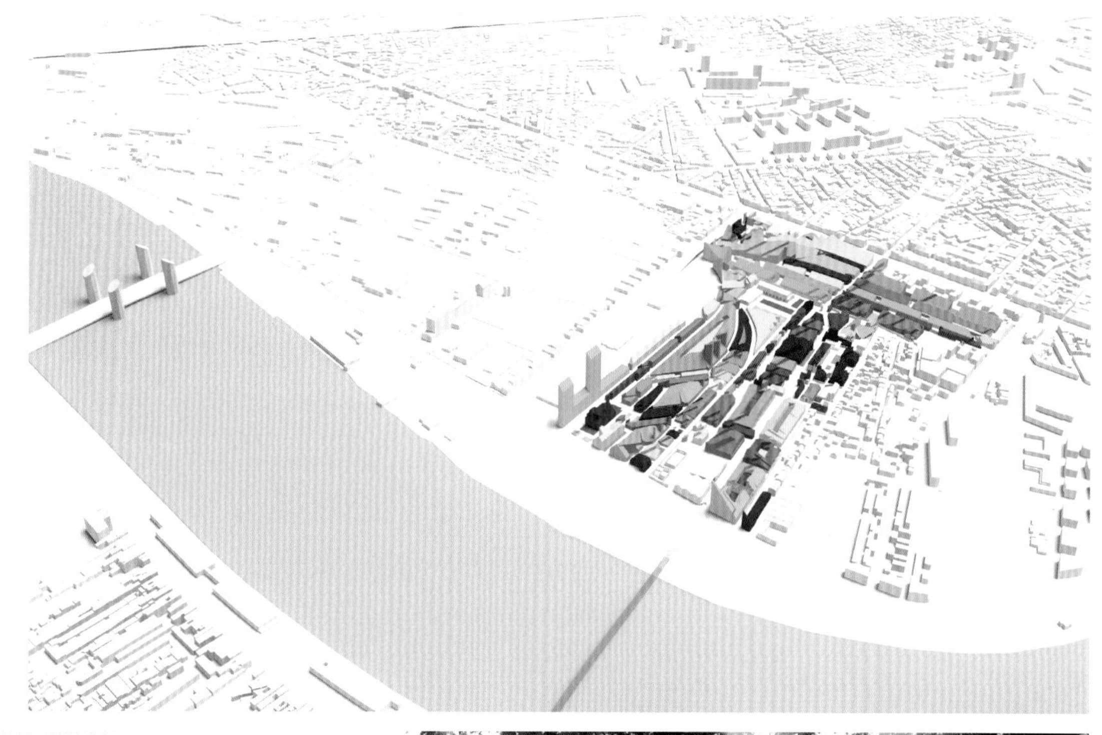

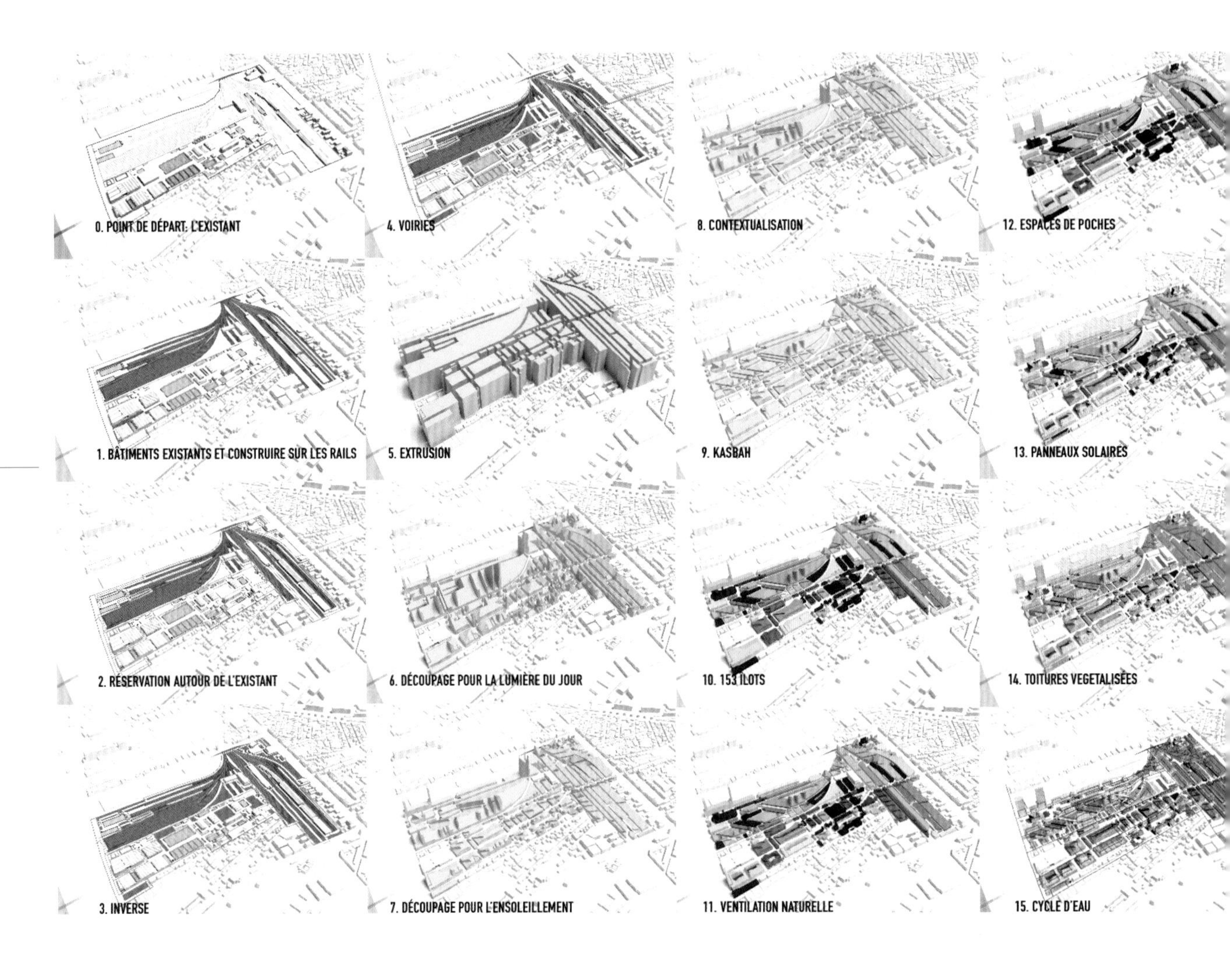
0. POINT DE DÉPART: L'EXISTANT
4. VOIRIES
8. CONTEXTUALISATION
12. ESPACES DE POCHES
1. BÂTIMENTS EXISTANTS ET CONSTRUIRE SUR LES RAILS
5. EXTRUSION
9. KASBAH
13. PANNEAUX SOLAIRES
2. RÉSERVATION AUTOUR DE L'EXISTANT
6. DÉCOUPAGE POUR LA LUMIÈRE DU JOUR
10. 153 ILOTS
14. TOITURES VEGETALISÉES
3. INVERSE
7. DÉCOUPAGE POUR L'ENSOLEILLEMENT
11. VENTILATION NATURELLE
15. CYCLE D'EAU

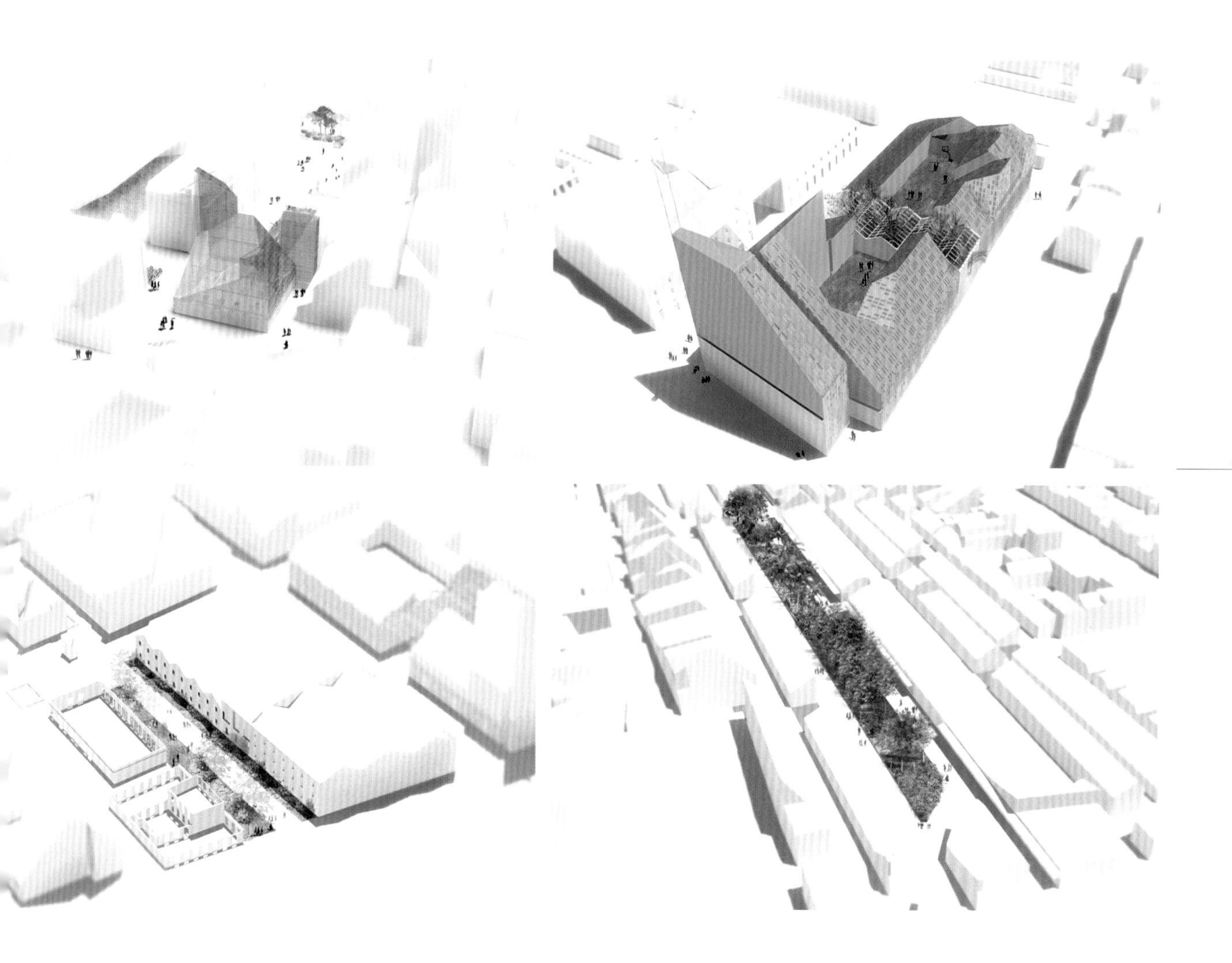

左图：体量研究
上图：长形口袋花园

Opposite: Volumetric representation
Above: The narrow pocket gardens

瑞士 纳沙泰尔 / 2011

Numaport
努玛波尔港

努玛波尔是纳沙泰尔湖泊上的港口，不仅犹如正面迎向阿尔卑斯山的一座舞台，也是呈现历史的场景。城市与湖泊之间具有亲近紧密的关系和源远流长的历史。纳沙泰尔城市与其湖岸之间的关系充满张力，为了城市发展而不断伸延湖岸界限。这个努玛波尔港口的项目也成为土地动态发展的一部分。

在此方案中，城市与湖泊的关系通过地面处理、新的植物栽种配置以及具有框景效果的构筑元素来达成。一系列紧密的城市空间逐一串联，并且以不同的空间层次朝向较为疏阔的岸边伸展。

这个组织空间的工作为城市历史带来崭新的阅读方式，在地与水之间设置的一些节点场所成为此港口街区整治的关键。让人们理解他们所处的土地并与其重新产生连结，是方案为了定义这些位于城市低处的整体空间所采取的行动。这份对土地的理解体现于不同时期出现的、地面向湖水延伸的行为上。

本方案荣获努玛波尔港口整治之设计竞赛第2名。

The Numaport, port of Neuchâtel embodies a stage which features both the Alps panorama and a more intimate and enduring duet between the city and the lake. Neuchâtel has always played the push and pull game with its coastline, extending its boundaries for further development. The Numaport project is part of a large regional ambition.

The connection to the lake is achieved through the ground, with the careful positioning of plantings and the erection of built architectural elements drawing the eye to the lake horizons. The combination of dense and intimate city areas and more open spaces towards the banks, extends over several levels down to the shore.

This composition work introduces a fresh interpretation of the city's history, by highlighting the meeting of land and water, acting as cornerstones in the port's neighbourhood planning. A unique understanding of the territory and harmony between men and the ground they live on are the founding principles of the project, the nature of which can be seen through the reclamation of land over water.

Runner-up in the Numaport competition.

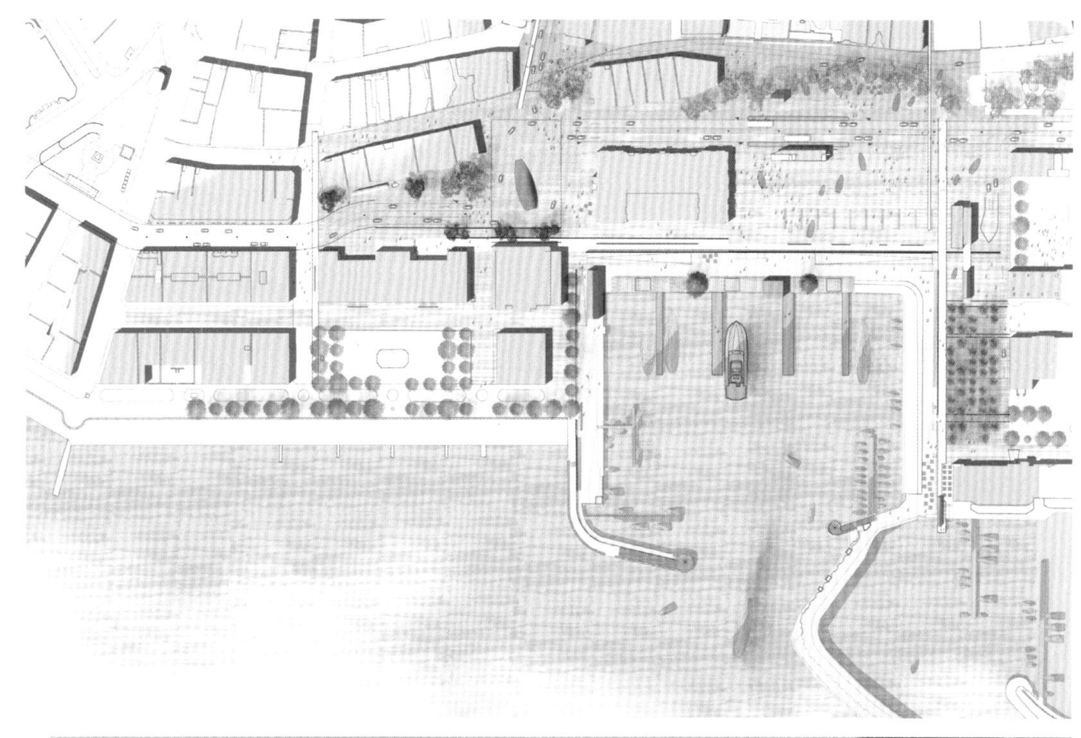

努玛波尔港广场，犹如一扇展现恢宏城市景观的窗户

Numaport square, a window to the unique city landscape

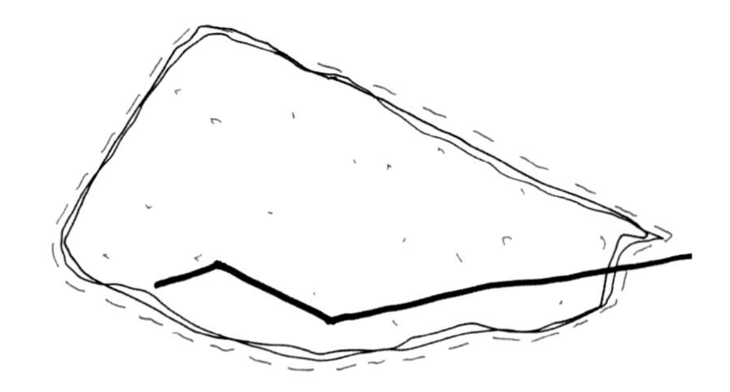

空旷之歌

embracing the void

法国 塞尔芒吉 / 2008

large meadow
大 草 原

位于汝拉山区的塞尔芒吉村庄围绕着一个名为“大草原”的广阔草地而发展。此空间长期以来作为市集用地，并且汇集了村庄的一部分径流，在草地的低处设有一个喷泉。

大地景观事务所（TERRITOIRES）在方案构思过程中所提出的所有问题最终导向一个简明清晰的答案：以谨慎适度的方法整治空间、以精致微妙的手法显露基地特色。

大草原的边界借助一些建造体而划定，而中央则保持空旷以提供各种村庄活动使用。其边缘处理，一部分种植成排的梣树，一部分则设置了与地势巧妙结合的步道，衬托出微微起伏的地形。木板步道为“大草原”北边划出界限，也提供人们一种穿越村庄的新方式。

本方案荣获2010年城市空间整治奖。

The village of Sermange, located in the Jura, is built around an extensive lawned area, called the Large Meadow. This area, which has long been used for markets, drains the rainwater from a part of the village. A fountain stands at the site's lower portion.

Throughout TERRITOIRES' design philosophy, one of the agency's mottos has always been plain and simple: measured intervention and subtle discovery.

The Large Meadow's layout is a central void, open to a range of communal activities. The structures on its outer edges are partly built along rows of ash trees, and partly highlighted by a wooden path following the subtle rhythms of its undulated topography. This wooden path closes the northern side, and offers an alternative way through the village.

Urban Planning Prize 2010

上图：方案设计细心地环绕着空旷草地来划定界限，以便赋予更好的空间品质与用途
左图：既存空间在尊重之中获得永恒

Above: Drawing boundaries around empty space to form its identity
Opposite: Existing features are carefully preserved

连续性
continuity

法国 当皮埃尔 / 2002-2007

footbridge, parking and square
桥道、停车场与广场

当皮埃尔村庄在2002年与2007年之间将三个项目委托给大地景观事务所（TERRITOIRES）进行设计。村庄的高度信任使得事务所的方案能够自由发挥，并且达到令人满意、符合现代精神的结果，同时得以在这些项目之间贯彻良好的城市空间逻辑。这三个方案联合塑造出一条带领人们从高原到平原、从城市到河流的路线。

村公所广场的设计重新使居民与他们的萨隆河建立起关系。此广场利用一系列的平台与阶梯，发展出与河水接触的崭新方式。平台供钓鱼者使用，而沉入河水的阶梯则供儿童戏水。此广场在冬季会局部淹水，见证了河水的起伏多变。

新设计的桥道不仅联系河流两岸，同时也为河流景观塑造出框景效果。这道悬空的桥道散发出某种程度的惊险感，使得过桥行为仿佛重新成为英勇之举。位于台地上的停车场配置着树木与提供学生们休息的长凳，当车子消失时，便化身为其他用途的空间。此三部曲空间犹如城市地形断面的样条，呈现出村庄的地理特征。

本方案荣获城市整治与建筑奖省级优胜榜。

From 2002 to 2007, the town of Dampierre entrusted three projects to TERRITOIRES. This relationship allowed them to collectively develop unrestricted, bold and modern projects, establishing a strong and appealing urban coherence, drawing an urban route from plateau to valley and from city to river.

The development of the town hall square redefines exchanges with the river Salon and its people through the succession of balconies and levels within the site, encouraging new ways of relating to the water. The balcony welcomes fishermen, while the stairs moistened by the waves delight the children. The site which is partly flooded in winter, reflects the river's changing levels.

The footbridge connecting the two banks frames the river landscape, displaying its delicate nature. Crossing the water becomes something of a thrill. On the plateau, a parking enhanced with trees and a long bench for schoolchildren, is transformed as the cars disappear. The triptych is an urban transect, uncovering geography.

Departmental Architecture and Urban Planning Prize

上图：方案设计创造出使萨隆河进入广场内部的可能性
左图：桥道（1）、停车场（2）与村公所广场（3 & 4）

Above: The river is given a free rein to take over the space
Opposite: Footbridge (1) parking (2) and town hall square (3 & 4)

恒常
consistency

法国 马赛 / 2003

Capucins long square
嘉布遣会修士长形广场

马赛的嘉布遣会修士长街从圣查尔斯火车站台地一直延伸到旧港口街区，其狭窄宽度为城市景观提供了绝佳的框景效果，让人完全领会这个城市的美感。它所穿越的历史街区也正是城市居民生活的最佳写照之一。

这个城市旧街区正经历着现代化的转变：拆除和重建的场面不断上演。大地景观事务所（TERRITOIRES）的设计方案提出一种反潮流做法，并且企图达到某种程度的一致性。将嘉布遣会修士长街改造为嘉布遣会修士长形广场成为这个场所历史的新里程，在此，唯有空间的用途必须经过现代化的调整。一条独特的步行轴线因应而生，并且在地形坡度明显向城市中心倾斜的位置点缀了一个较为宽阔的空间，让人们得以眺望城市和远处丘陵。

方案设计刻意让这个空间成为不论是对造访过客或对马赛人本身而言，都极为率直和简约的场所。整个长形广场都以相同的设施来呈现嘉布遣会修士的行迹以及居民对街区的情感：石板地面让人想起昔日旧城，而凌空横挂的轻纱则为景观视野带来框景效果。此地成为人们在这条长街上换气、休息的场所，也是一个观景平台，并且设置了一个喷泉和露天咖啡座。这个长形广场纯然展现着地中海气息，以其独特的地面、光纤和消磨时光的方式，来与城市的尘埃共舞。

The Capucins long street stretches from the Saint-Charles railway station plateau to the old port of Marseille. Its narrow silhouette and exceptional layout through the cityscape displays the beauty of the town. Cutting through its historic district, and subtly picking up the rhythm of its residents' activities.

This area favours modernity through demolition and reconstruction. TERRITOIRES' proposal goes against the norm, aiming for a certain consistency. Converting the long street into a long square provides a new approach to the site's history, where modernisation is contained within its practical uses. From there a particular pedestrian axis, expanding in length in the plunging centre of Marseille, provides transient views to the hills.

This area strives for simplicity and sincerity for both local people, and those passing through. A strategy has been put in place along the route to evoke the old layout of the Capucins and its neighbourhood's commitment to the area. The paving alludes to the ancient city and sheer hanging drapes frame the views of the surrounding landscape. The area becomes a haven, offering respite on the hill, a panoramic viewpoint providing a niche for a fountain and café. The square is thereby simply defined as Mediterranean in its nature, its light and its attitude to the passing of time, a playful witness to the city's drifting dust.

natures

自 然

景观设计师不断见证着人类社会与自然环境之间的复杂关系，而大自然正是他们的创作原料。

大地景观事务所（TERRITOIRES）对此课题备为关注，并发展出一些实际行动，自诩成为那些“能够引导生命”之过程的摆渡人。怀着对大自然的颂扬之情，大地景观事务所将人们带领到结构简单却令人赞叹的设施中，让人类与自然彼此交融，不相互竞争也不彼此压迫。那些自然生物虽是微妙、脆弱和敏锐的，却也具有宏伟和壮观的特性。

Landscape architects use nature as their raw material, and have always attested to a complex relationship between human society and nature.

TERRITOIRES' involvement is sensitive to this issue and supports the processes that channel life. Through its extraordinary creations, TERRITOIRES inspires the respect of man and nature, by investing in them, without negating or competing with them. This choice aims to reflect the delicate, fragile and sensitive qualities of nature, as well as its great strength and constitution.

拟态
mimicry

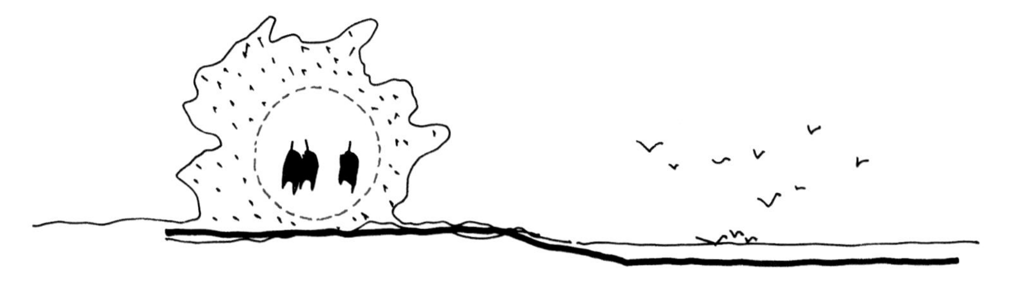

法国 孔吉-苏尔-碟环讷 / 2015

Grand Voyeux regional nature reserve
格宏德瓦佑地区级自然保护区

距离巴黎40千米的格宏德瓦佑地区级自然保护区是人们对马恩河进行砾石开采历时30年而塑造出的人为环境。对此地进行规划与设计是件非常棘手的事，不仅必须保护这个才勉强复原的景观，同时也要将基地广泛开放给大众使用。方案所提倡的发展方式与一般在自然环境常见的标准教学式整治颇为不同，为了建立低调的观察模式，于是决定以拟态掩饰的方式来融入环境状态中。

方案效仿长脚鹬行为，它在舒适的湿地环境中走动，却毫不声张虚事。整个参观路线以两种互补的设施来建立：轻盈的木板通道和柔和的绿色通道。木板通道延展于整个基地，遇到特殊环境则加设观望台。绿色通道则以当地植物为基础，强化植物效果也适时地将参观者隐藏起来。

观望台形如帆布蛹，设置于沙沙作响的芦苇丛中，提供给观察者意想不到的惊奇赞叹。这些观望台也使得访客能够躲开鸟禽视线，并且在舒适的环境中观察大自然里的动物。

The Grand Voyeux Regional Nature Reserve is a man-made park, 40 km from Paris, which has taken shape over thirty years with the quarrying of gravel pits on the banks of the Marne. This complex project's strategy distanced itself from the usual didactic development standards for natural environments, in a bid to recover a barely healed landscape, in order to open it up to a wider public. Camouflaging is required to ensure the readjustment of the landscape is discreet.

The project, inspired by stilts, enables comfortable mobility through wetlands without marking the ground. The route is built using two complementary systems: light wood decking and a vegetal tunnel. The decking is stretched in its general geometry and reinforced by strategic observatories. The pipe is supported by native vegetation, concealing hikers.

The observatories, chrysalises made of canvas, laid on the rustling phragmites, provide unexpected and captivating refuges for bird watchers. These shapes hide visitors from the birds, allowing animal watching in exceptional comfort.

保护区之家，提供游客第一个观赏基地的视野，也是整个参观路线的起点

The Maison de la Réserve provides the first view of the site and creates a gateway to the path

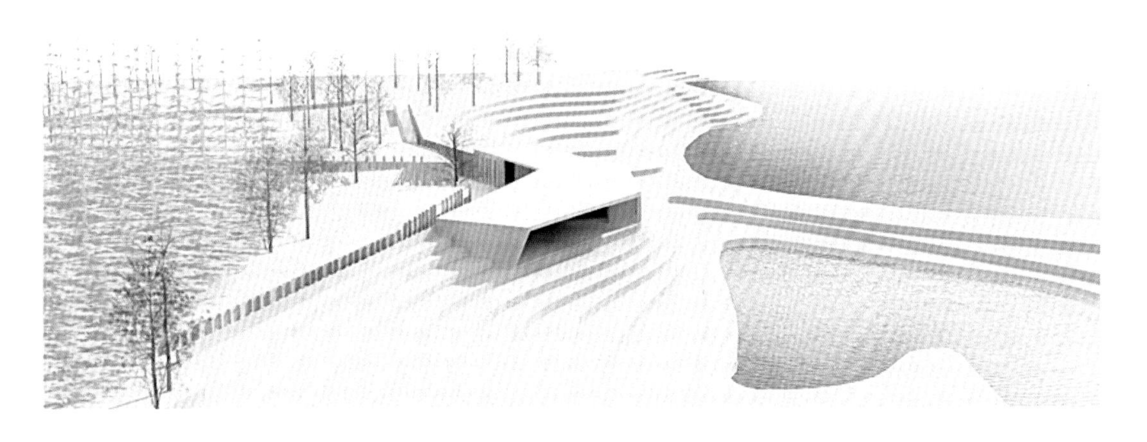

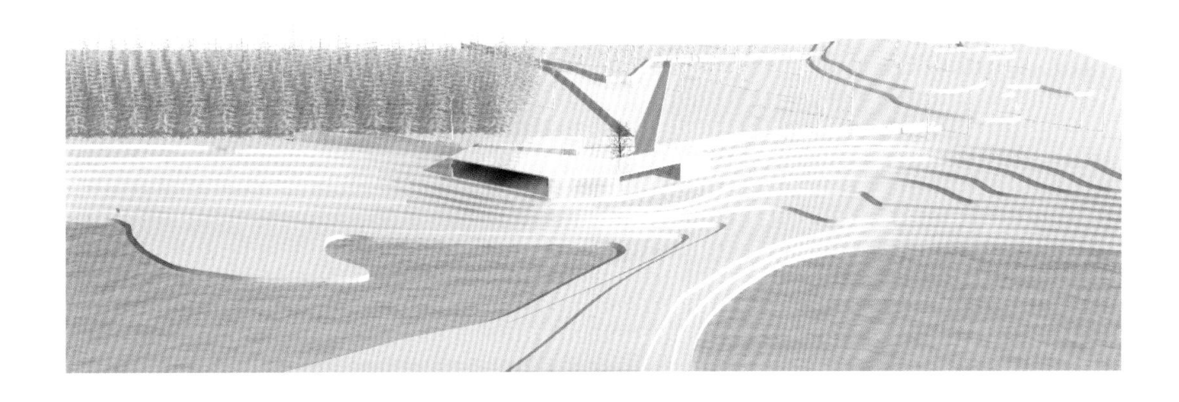

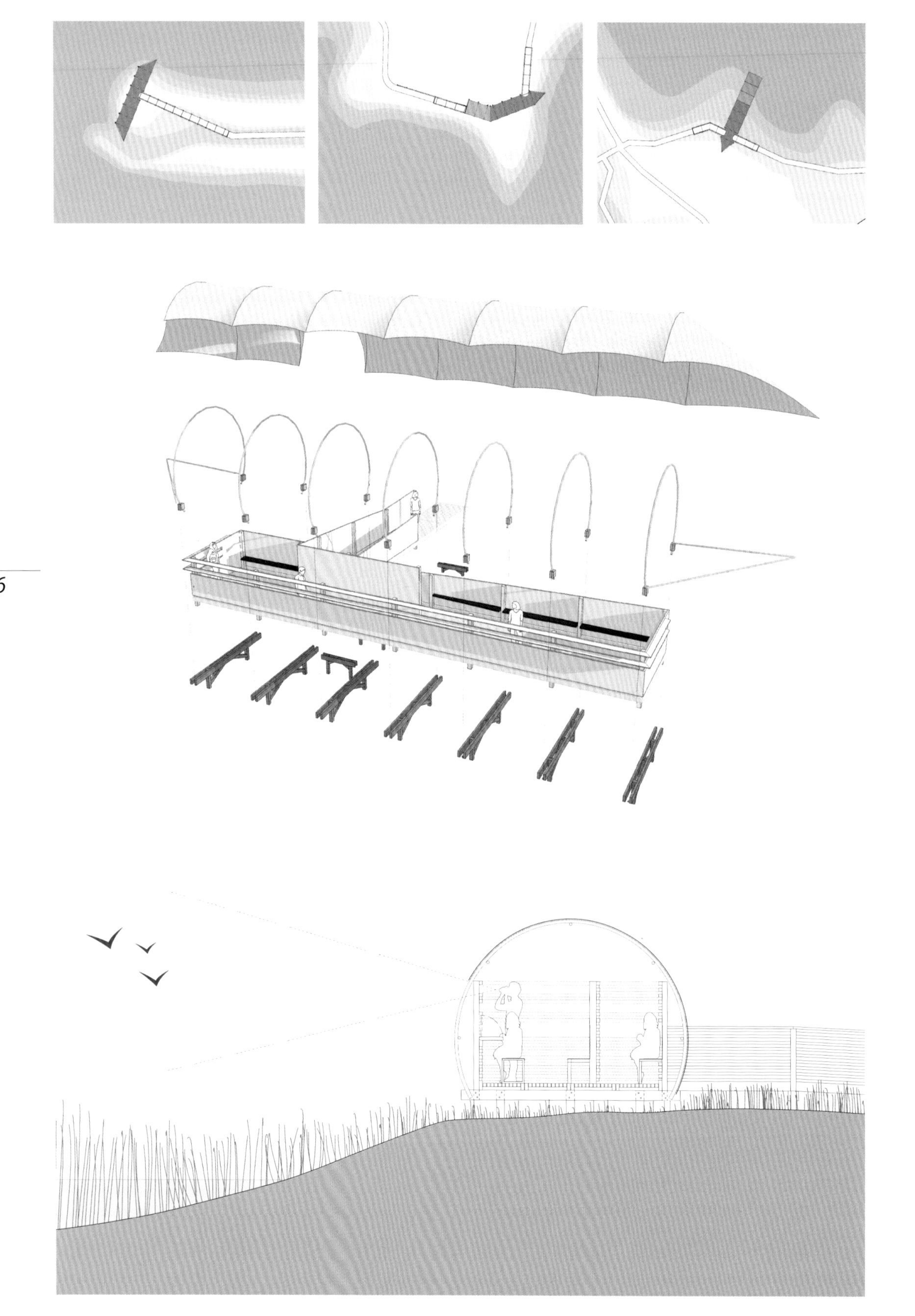

观望台提供参观者一个认识保护区的丰富自然景观的绝佳视点

Observation points showcase the reserve's natural riches

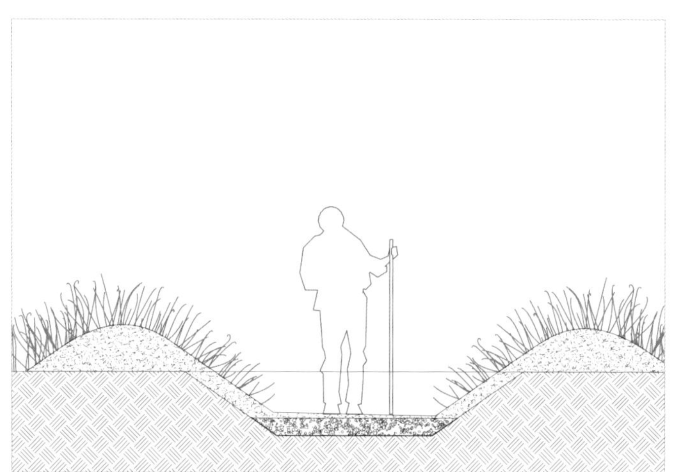

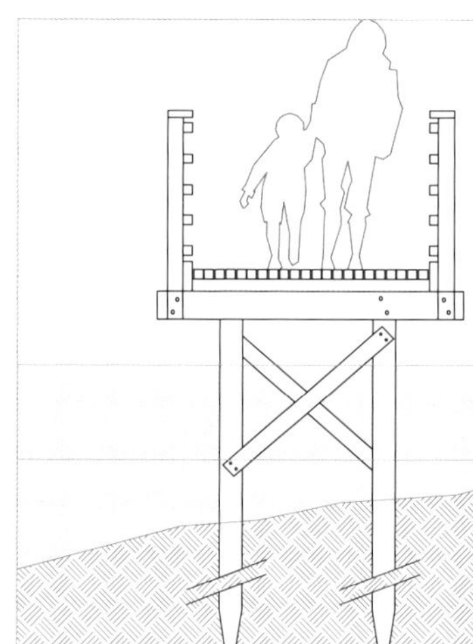

参观路径经过特别的设计考量，借助基地本身的植物来将访客隐藏起来

Movement is designed to ensure visitors are camouflaged by the site's vegetation

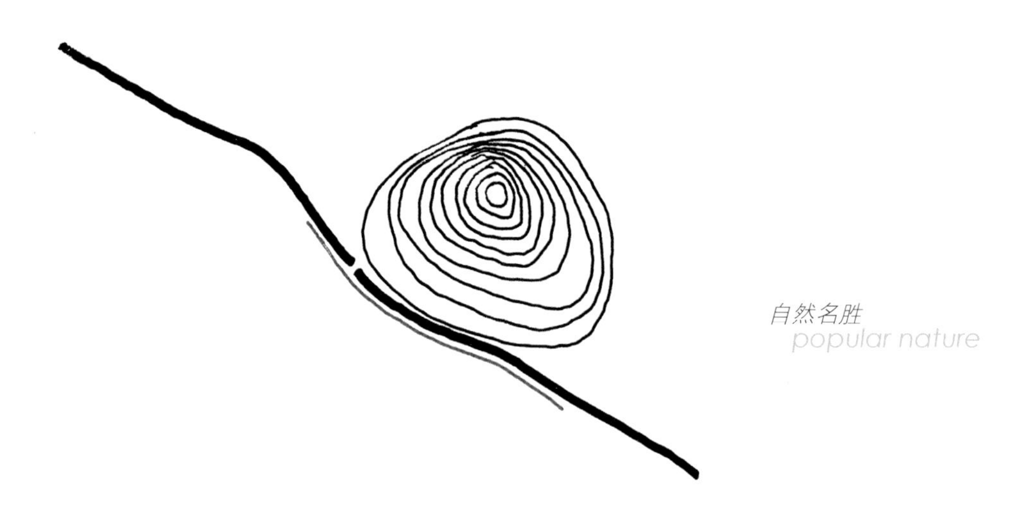

自然名胜
popular nature

法国 圣马尔夏勒&圣俄拉里 / 2015

Mont-Gerbier-de-Jonc tourist site
热尔比耶德荣山自然观光基地

热尔比耶德荣山犹如旺度山、皮拉沙丘和赫兹海岬等地，都属于法国人民集体意识的一部分，被列为“法国自然名胜”的保护基地，是观光客络绎不绝的旅游景点。然而，长期的开放观光却造成了基地容貌的改变，法国政府不得不对此基地的景观与用途重新进行省思。

本项目为此场所提供了新生的机会，使其基础元素得以复苏：地质、地理与热门旅游。就地质而言，火山山脉——热尔比耶德荣山是其中的突起元素——造就了这块土地的特征，特别是响岩圆锥，形成了此地的特殊景观。就地理而言，全法国数百万的学生们都会在地图上指出这个地点，因为这里是法国最大河流——卢瓦尔河——的天然起源地。

就热门旅游而言，早在1930年，通往这个景点的道路被修建之时，法国旅游俱乐部就已经可以预见这里成为观光胜地的情况。大地景观事务所（TERRITOIRES）的设计方案试图彰显这条风景非凡的道路，使它成为阅读大地景观的工具；同时，事务所也为道路的边缘做细致的处理，将其视为犹如介于自然与文化之间的珍贵雕饰，借以强化道路的角色。基地的几何与材质特征因而被重新显现出来。热尔比耶德荣山的整治项目必须在保护景观与符合演变需求之间取得平衡，同时也需要一定的自主性，以免落入看似幸福却缺乏思辨能力的自然主义结局。

The Mont-Gerbier-de-Jonc belongs to French collective consciousness as do the Mont Ventoux, the Dune du Pilat and the Pointe du Raz. Recognised as one of France's sites of outstanding natural beauty, this tourist hot-spot has been overpopular with visitors over the years, prompting a rethink of its landscape and uses.

The project is an opportunity to modernise the site's geology, geography and tourism. The chain of volcanic peaks – including the Mont-Gerbier-de-Jonc peak – has shaped the area, in particular the sugar-loaf phonolites dominating the landscape. Millions of the nation's schoolchildren have become familiar with this site's geography, the location of which is also home to the source of the Loire, the longest river in France.

In the 1930s, the Touring Club de France anticipated the site's popularity with the construction of an access road to the area. TERRITOIRES' project enhances this scenic road, restoring its role as an interpretation tool, reaffirming it with precise boundaries, a subtle illustration of the relationship between nature and culture. The site's geometry and materiality are revived. The planning of the Mont-Gerbier-de-Jonc is a delicate development involving the conservation as well as the necessary evolution of the area, which aims to avoid an inflated sense of naturalism.

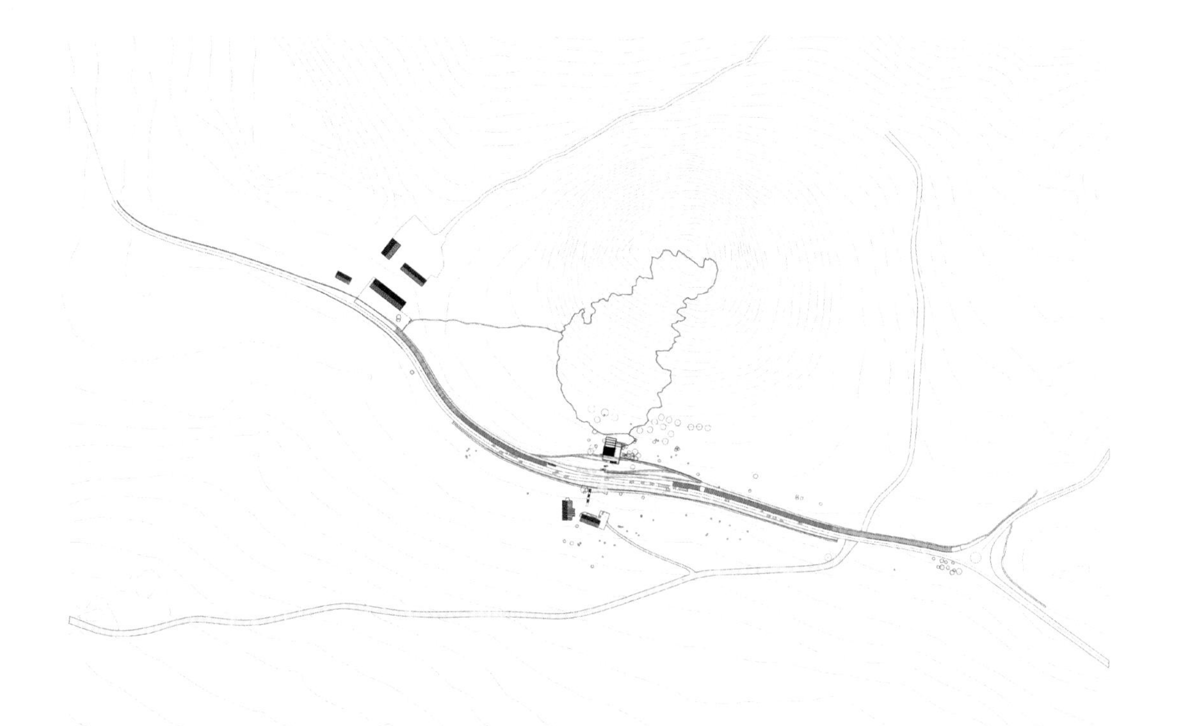

方案在热门旅游的需求与地质遗产的保护之间取得综合性的平衡

The project synthesises popular tourism and its geological heritage

法国 诺尔热–拉–维勒 / 2013

wet meadows and source of the river Norges
诺 尔 热 河 发 源 地 与 湿 地 草 原

位于第戎北边、朗格勒高原脚下的诺尔热–拉–维勒村庄包含了两个沿着诺尔热河发展的聚落，这条河流的发源地便在此处。然而，河流经常性的泛滥使得居民无法享受此地的旷野景观之美。

大地景观事务所（TERRITOIRES）为此项目选择了一个既根本又极简的设施：一条细致平薄的木板步道。这条步道延伸于两个聚落之间，建构出绵长的远景，它不时与河流亲近调情，并且在路线中点处以过桥形式穿河而过。河流的蜿蜒曲线和步道的笔直线条形成强烈对比，使在其中散步的人们仿佛身处于惊奇绝妙的悬浮状态，双脚从未碰到真实的地面。

此方案借助全程贯彻的美学元素和极简的人为介入，使人们重新找回此处的大地景观。

The town of Norges-la-Ville to the north of Dijon, at the foot of the Langres plateau, is spread across two villages on the Norges. This is where the rvier takes its source from. Frequent floods have deprived the locals of an access to the wild and beautiful landscape.

For this project, TERRITOIRES chose the radical and minimal system of a wooden path, which is as smooth as glass. This path extends across both villages, frequently teasing and eventually crossing the river halfway via a footbridge. The geometric contrast between the sinuous river and the path's rigid lines give its visitors an astonishing sense of levitation, without ever touching the ground.

This landscape is rediscovered through a radical and uncompromising aesthetic, with minimal human intrusion.

精心安排的参观路线让人们与河流产生动人的亲密关系，并借此认识河畔的环境与景观

The river setting invites exploration via an intimate path

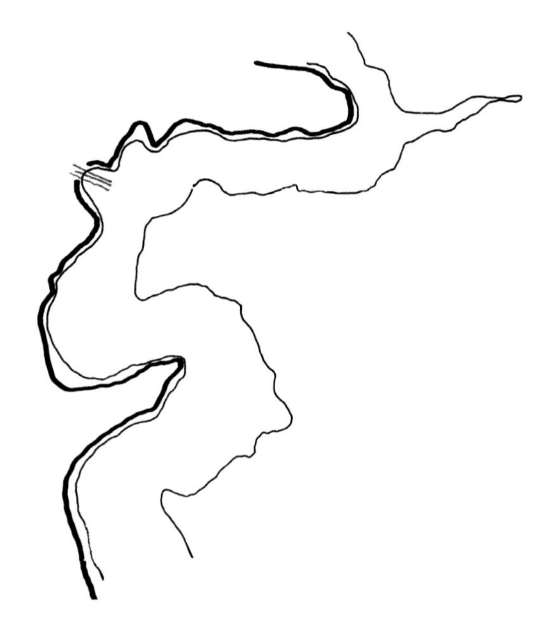

土质崩解
slaking

法国 乌格朗 / 1998

marina
游 艇 码 头

法国政府于1965年决定在艾因河上建设一个水库。安德烈·贝松在其小说《Le Village englouti》（淹没的村落）当中提到两个如今已经沉浸于水中的农村，而沃克吕兹省的查尔特勒修道院也遭到同样的命运。30年后，在河左岸建设一个游艇码头再度为此地的景观改写历史，不仅创造出休闲娱乐的氛围，也建立一种与大自然之间较为轻松的关系。

方案从一个简明的计划书内容出发：建造可停靠300艘游艇的三座浮桥。这些浮桥的笔直线条与河岸柔和曲线的结合，形成一个清晰易读的空间，使游船者与他们的船只、当地植物与它们所处的环境，皆各得其所。

这三条木质直线伴随着由柳树和芦苇所组成的不同植物密度，提供人们从陆地过渡到水面的经验。访客的步行路线必须因应水坝运作所造成的多变水位而设置，其中某些路径能够承受长时间沉浸于水中，某些则以漂浮形式存在，并随着水位下降而毫无阻碍地附着地面。这些设计考量显现出方案在面对项目所处环境时，所持有的细心关注的态度。

In 1965, the government decided to build a reservoir on the Ain. In his novel "Le Village Englouti" (The Sunken Village), André Besson evokes the memory of two farming towns which are now submerged. The Chartreuse de Vaucluse suffered the same fate… Thirty years later, the construction of a marina on the left bank takes this landscape on another journey; one of leisure and a more relaxed approach to nature…

The project adopts a sober and simple approach, with three pontoons accessible to 300 boats. The contrast between the rigid geometry of these lines and the bank's more flexible design creates a coherent space suited to all its users, yachtsmen, their boats, the local wildlife and its environment alike. Three wooden bands create a transition from land to water through a succession of layers of vegetation, including willows and phragmites.

Pedestrian management systems adapt to significant changes in the lake level, subject to the dam's activity. Some pathways are able to withstand prolonged immersion, others are equipped to float and settle safely. The project is responsive to its environment.

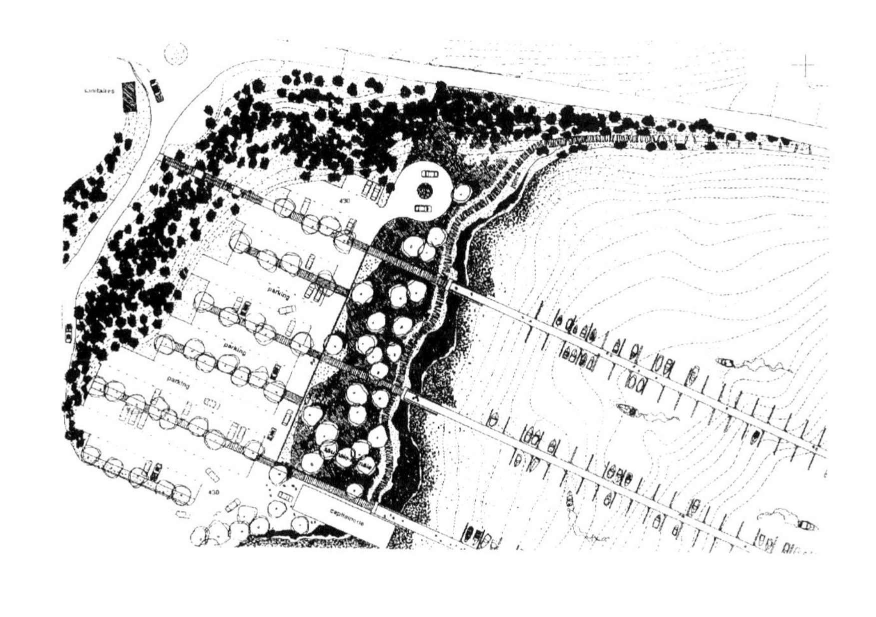
sanitaires
430
parking
parking
parking
430
capitainerie

停车场、交通流线和船只泊位的组织皆围绕着河岸与水面之间的三条直线而展开

Parking, circulation and boat mooring are structured along the lines extended between the shore and water

projects index

方案索引

map of featured projects
本书项目地理分布图

(P) : 已建造或正在建造之方案 Current or ongoing project
(E) : 已完成或正在进行之研究 Current or ongoing study
(C) : 竞赛 Competition

花园 gardens

2016

克莱尔方丹庄园 **CLAIREFONTAINE ESTATE** (P) pp.14-17
Clairefontaine-en-Yvelines, France
设计者 / By : BMT & Associés Architectes (project representative), Atelier Terrible (project representative), Territoires, Etico, Huguet, Venathec, Mediaplex
业主 / For : CTNFS de Clairefontaine
12.3 M€ ex. VAT

2013

BÂTIE WOOD PARK (C)
Geneva, Switzerland

LA CENTRALITÉ ZAC (COMPREHENSIVE DEVOLPMENT ZONE) URBAN ECO PARK (C)
Carrières-sous-Poissy, France

2012

DONNER PARK (C)
Montreux, Switzerland

2011

塞尔日·甘斯布花园 **SERGE GAINSBOURG GARDEN** (P) pp.18-23
Paris, France
设计者 / By : Territoires (project representative), Gelin & Lafon, Light Cibles, ATPI, Integral 4
Pour / For : SEMAVIP
10.5 M€ ex. VAT

贝尔福-蒙贝利亚尔高速火车新车站
BELFORT-MONTBÉLIARD NEW TGV STATION (P) pp.24-29
Meroux, France
设计者 / By : DAAB (project representative), AREP (project representative), Territoires, D'Ascia, RFR, Le point Lumineux, HAEFELI
业主 / For : SNCF (Stations and Escale Branch)
12.8 M€ ex. VAT

2010

GLACIS PANORAMIC VIEWPOINT (C)
Besançon, France. Award winners

PARC DU REPOSOIR REPOSOIR PARK (C)
Nyon, Switzerland

2009

瓦德贝巴公园 **VALDEBEBAS PARK** (C) pp.30-33
Madrid, Spain
设计者 / By : Territoires (project representative), NA Arquitectura
业主 / For : Madrid City Council

PLANOISE URBAN PARK (C)
Besançon, France

2006

LIVERY STABLE PARK SITE (C)
Le Mans, France

COLLEGE (C),
Burnthaupt, France

2005

MONTÉVRAIN PARK (C)
Montévrain, France

TOWN CENTRE (P)
Montlebon, France

TREFFEL GARDEN (P)
Vandœuvre-lès-Nancy, France

SCHOOL OF AGRICULTURE AND HORTICULTURE (C)
Saint-Germain-en-Laye, France

2003

椴树广场 **TILLEULS SQUARE** (P) pp. 34-39
Gray, France
设计者 / By : Territoires (project representative)
业主 / For : Commune de Gray
960 000 € ex. VAT
荣获城市空间整治奖，建筑奖省级优胜榜 / Urban Planning Award, Regional Architecture Award

CITY HALL GARDEN (P)
Oyonnax, France

LAMUGNIÈRE PARK (P)
Arc-lès-Gray, France

2001

CULTURAL CENTRE GARDEN (P)
Boult, France

CHAMBER OF AGRICULTURE TRAINING INSTITUTE PARK (P)
Saint-Quentin-en-Yvelines, France

LOCHY PARK (C)
Magny-le-Hongre, France

1999

EQUESTRIAN CENTRE (C)
Draveil, France

历史 histories

2011

一战博物馆 **MUSEUM OF THE GREAT WAR** (P) pp.42-45
Meaux, France
设计者 / By : Atelier Lab (project representative), Territoires, Greusat
业主 / For : Agglomeration Community of Pays de Meaux
3.9 M€ ex. VAT

2010

英国人散步大道 **PROMENADE DES ANGLAIS** (C) pp.46-51
Nice, France
设计者 / By : Rudy Ricciotti Architecte (project representative), Territoires, Le point Lumineux, Pascaline Minella
业主 / For : Nice City Council
120 M€ ex. VAT
设计竞赛结果保留 / Competition withheld

2009

MUSIC CONSERVATORY (C)
Villemomble, France

2007

FORT BARRAUX CONVERSION (E)
Feasibility study
Barraux, France

2006

镇公所花园 **TOWN HALL GARDEN** (P) pp.52-57
Sceaux, France
设计者 / By : Territoires (project representative), Kahane, Inex, Yves Morel, Tone
业主 / For : Sceaux Town Council
1.06 M€ ex. VAT

圣阿尔宾运河 **SAINT-ALBIN CANAL** (E) pp.58-59
Scey-sur-Saône-et-Saint-Albin, France
Par / By : Territoires (project representative), Pascaline Minella
业主 / For : DRAC de Franche-Comté, Voix Navigables de France
44 000 € ex. VAT

沙畹洗煤厂 **CHAVANNE'S COAL WASHING PLANT** (E) pp.60-63
Montceau-les-Mines, France
设计者 / By : Agence MVRDV (project representative), Territoires, Ferraru, Alto, BPI, Coyne & Belier, Vanguard
业主 / For : Community of Creusot-Montceau
108 000 € ex. VAT

2005

CULTURAL CENTRE (C)
La Seyne sur Mer, France, Architect ANMA

GOBERT RESERVOIRS (C)
Versailles, France, Architect ANMA

2004

MUSEUM OF FINE ARTS (C)
Angers, France

ECOMUSEUM OF THE CHERRY (P)
Fougerolles, France
Rigional Architecture Award

1998

FORTRESS ZOOLOGICAL GARDENS (C)
Besançon, France

JURA ARCHEOLOGICAL MUSEUM (C)
Lons-le-Saunier, France
Award winner (abandoned project)

CATHEDRAL SQUARE (C)
Sens, France

1995

沃邦城堡水族馆 **VAUBAN CITADEL'S AQUARIUM** (P) pp.64-67
Besançon, France
设计者 / By : Territoires (project representative), Quirot & Vichard
业主 / For : SEM de la Citadelle
76 000 € ex. VAT

城市 cities

2018

果园生态小区 **LES VERGERS ECO-NEIGHBOURHOOD** (P) pp. 70-73
Meyrin, Switzerland
设计者 / By : Territoires (project representative), Amsler-Bombeli, Farra & Zoumboulakis, ON, Philippe Cabane, BCPH
业主 / For : Meyrin Town Council
30 M€ ex. VAT

2017

SALINS-LES-BAINS NEW THERMAL BATHS (P)
Salins-les-Bains, France

2014

内贝尔花园 **NEPPERT GARDENS** (P) pp.74-77
Mulhouse, France
设计者 / By : Agence Nicolas Michelin et Associés (project representative), Territoires, OGI, ON
业主 / For : SERM
11.2 M€ ex. VAT

AUXON-DESSUS ZAC (COMPREHENSIVE DEVOLPMENT ZONE) (P)
Auxon-Dessus, France

SCHOOL COMPLEX (P)
Attigny, France

CULTURAL CENTRE AND MULTI-MEDIA LIBRARY (P)
Giromagny, France

2013

SAINT PANTALÉON NEIGHBOURHOOD (C)
Autun, France

NORTH-SOUTH STATION NEIGHBOURHOOD LINK (E)
Béthune, France

ELIOR MOTORWAY SERVICES (C)
Le Mans, France

CONCORDE NEIGHBOURHOOD (C)
Geneva, Switzerland

KIEM NEIGHBOURHOOD (C)
Luxembourg, Luxembourg

2012

CEVA-CHAMPEL HOSPITAL STOP (C)
Geneva, Switzerland

尼尔城堡 **BASTIDE NIEL** (E) pp.78-81
Bordeaux, France
设计者 / By : Agence MVRDV (project representative), Territoires, Oasiis, Arcadis, Davis Langdon
业主 / For : Bordeaux Urban Community
34.5 M€ ex. VAT

NEW THERMAL BATHS (P)
Salins-les-Bains, France

URBAN DEVELOPMENT PLANNING PROGRAMME (E)
Salins-les-Bains, France

2011

努玛波尔港 **NUMAPORT** (C) pp.82-85
Neuchâtel, Switzerland
设计者 / By : Territoires (project representative), B+S Ingénieurs, BCPH, Studio Vicarini, Damien Cabiron
业主 / For : Neuchâtel Town Council
14 M€ ex. VAT
努玛波尔港口公共空间整治设计竞赛第2名 / Runner-up in the Numaport Competition for Public Spaces

CEVA CAROUGE-BACHET STOP (C)
Geneva, Switzerland
Runner-up in the Competition for Public Spaces

PENINSULA DEVELOPMENT (E)
Caen, France, Architect MVRDV
Project definition

COTEAUX NEIGHBOURHOOD (E)
Mulhouse, France
Green landscape study

LGV NORTH BANK FOR THE BELFORT TGV STATION (E)
Meroux, France
Landscape study

STATION NEIGHBOURHOOD (C)
Sion, Switzerland

MANTES UNIVERSITY NEIGHBOURHOOD (P)
Mantes-la-Jolie, France, Architect Fortier

SCHOOL COMPLEX (P)
Attigny, France

THE BARRACKS MULTICULTURAL CENTRE (P)
Giromagny, France

STATION (C)
Sion, Switzerland

PETIT SACONNEX SQUARE (C)
Genena, Switzerland

2010

UNIVERSITY CAMPUS (E)
Montpellier, France, Architect MVRDV
U&U study

LES COURTILLES AREA REDEVELOPMENT (E)
Asnières, France
Project definition

METZ NEIGHBOURHOOD - THE WORLD'S END (E)
Beurre, France
Landscape study

AUSTERLITZ SQUARE (C)
Strasbourg, France

DESCARTES NEIGHBOURHOOD UNIVERSITY CAMPUS (C)
Marne-la-Vallée, France, Architect MVRDV

FAISCEAU URBAN PROJECT (C)
Nanterre, France, Architect MVRDV

2009

PITRES MULTIMODAL PORT PLATFORM (E)
Le Manoir Alizay, France
Feasibility study

2008

大草原 **LARGE MEADOW** (P) pp.86-91
Sermange, France
设计者/ By : Territoires (project representative)
业主 / For : Sermange Town Council
245 000 € ex. VAT
2010年城市空间整治奖 / Urban Planning Award 2010

CHURCH, TOWN HALL AND MARKET SQUARE (C)
Saint-Yorre, France

2002-2007

桥道、停车场与广场 **FOOTBRIDGE, PARKING AND SQUARE** (P) pp.92-95
Dampierre-sur-Salon, France
设计者 / By : Territoires (project representative)
业主 / For : Dampierre-sur-Salon Town Council
229 000 € ex. VAT
城市整治与建筑奖省级优胜榜 / Regional Architecture and Urban Planning Prize

2007

PRÉS DE VAUX INDUSTRIAL SITE CONVERSION (E)
Besançon, France, Architect SEURA
Project definition

NORTH COMMERCIAL AREA PLANNING (E)
Strasbourg, France, Architect SEURA
Project definition

2006

GREEN AND BLUE NETWORK (E)
Montbéliard, France
Project definition

LIVERY STABLE PARK SITE (C)
Le Mans, France

BIÈVRE SYMBOLIC ITINERARY (E)
Paris, France
Project definition

COAL-BASED WASH HOUSE CHAVANNES SITE CONVERSION (E)
Montceau-les-Mines, France, Architect MVRDV
Feasibility study

UNIVERSITY NEIGHBOURHOOD – INNOVAPARC (E)
Mantes-la-Jolie, France, Architect B. Fortier
Project definition

PASTEUR SQUARE (C)
Besançon, France, Architect SEURA

HAUTS DU CHAZAL (C)
Besançon, France

BOCKSTAEL SCHOOL (C)
Brussels, Belgium

2005

VAUBAN NEPPERT NEIGHBOURHOOD (E)
Mulhouse, France, Architect Nicolas Michelin ANMA
Project definition

LEREBOURG WASTELAND REDEVELOPMENT (E)
Liverdun, France
Project definition

DEVELOPMENT OF THE RIGHT BANK OF THE MOSELLE (E)
Thionville, France
Project definition

SOUTH URBAN AREA (E)
Grenoble, France, Architect Nicolas Michelin ANMA
Project definition

"ONE CITY, TWO STATIONS, WHICH URBAN PLANNING PROJECTS ?" (E)
Dijon, France

RIGHT BANK PLAINS (E)
Nancy, France
Project definition

CULTURAL CENTRE (C)
La Seyne-sur-Mer, France, Architect Nicolas Michelin ANMA

FAGET SQUARE GARDENS (P)
Brussels, Belgium

HORTICULTURAL SCHOOL (C)
Saint-Germain-en-Laye, France

ESPOIR HOUSING ESTATE (P)
Montreuil, France

TOWN HALL SQUARE AND ITS SURROUNDINGS (C)
Monéteau, France

2004

ZPPAUP (PROTECTION OF ARCHITECTURAL, URBAN, AND LANDSCAPE HERITAGE) LANDSCAPE COMPONENT (E)
Grenoble, France, Architect G. Balduini

EDUCATIONAL CENTRE (P)
Boult, France

MARKET SQUARE (C)
Beauvais, France, Architect Nicolas Michelin ANMA

INCINERATION PLANT (P)
Carrières-sur-Seine, France

COMMUNITY CENTRE (P)
Courtefontaine, France

2003

嘉布遣会修士长形广场 **CAPUCINS LONG SQUARE** (C) pp.96-99
Marseille, France
设计者 / By : Averous & Simay (project representative), Territoires, AAHJ
业主 / For : Euroméditerranée
735 000 € ex. VAT

QUATERNAIRE PARK (E)
Haute-Loire, France
Feasibility study

DARSE NORD (P)
Macon, France

REDEVELOPMENT OF PUBLIC SPACES (E)
Raon-l'Étape, France
Project definition

FONTENEILLES CONVERSION AND DEVELOPMENT PROJECT (E)
Beaucourt, France

COLLEGE (P)
Fessenheim, France

SQUARE AND FOUNTAIN (P)
Chalezeule, France
Regional Architecture Award

2002

PARK AND WATER STATION DEVELOPMENT (E)
Besançon, France
Feasibility study

COLLÈGE COLLEGE
Marcolsheim, France (C)

LANDSCAPE ANALYSIS AS PART OF THE LOCAL TOWN PLANNING REVIEW (E)
Evette Salbert, France

LANDSCAPE ANALYSIS AS PART OF THE LOCAL TOWN PLANNING REVIEW (E)
Beaucourt A.U.T.B, France

OFFEMONT A.U.T.B. (E)
Offemont, France
Feasibility study

COMMUNITY CENTRE (P)
Saône, France

2001

LANDSCAPING FOR THE TGV STATION LINK (E)
Besançon-Auxon, France

STONE GUTTER RESTORATION (P)
Mont-Saint-Léger, France

SCHOOL APPROACH (C)
Ornans, France

PONT DU ROUTOIR NEIGHBOURHOOD (P)
Guyancourt, France

2000

REVISION OF THE LAND USE PLANNING (E)
Montbéliard, France

BOULEVARD MARÉCHAL FOCH PLANNING (E)
Le Creusot, France
Preliminary design study

TRAM TERMINAL (C)
Nancy, France

ZPPAUP (PROTECTION OF ARCHITECTURAL, URBAN, AND LANDSCAPE HERITAGE) (E)
Montbéliard, France

BOULEVARD DE LA RÉPUBLIQUE (P)
Châlons-sur-Saône, France

OLD CENTRE (P)
Sassenay, France

SCHOOLHOUSE (P)
Fédry, France

1998

CHARLES LE TÉMÉRAIRE SQUARE (P)
La Rivière-Drugeon, France

PLACE DE LA LIBERTÉ (C)
Lons-le-Saunier, France

HOUSE OF DANCE (C)
Lyon, France

DAUPHINÉ GROUNDS (C)
Lyon, France

自然 natures

2015

格宏德瓦佑地区级自然保护区X
GRAND VOYEUX REGIONAL NATURE RESERVE (P) pp.102-109
Congis-sur-Thérouanne, France
设计者 / By : Territoires (project representative), Charles-Henri Tachon, ATPI, Integral 4, OGE
业主 / For : AEV Paris
1.7 M€ ex. VAT

热尔比耶德荣山自然观光基地
MONT-GERBIER-DE-JONC TOURIST SITE (P) pp.110-113
Saint Martial & Sainte Eulalie, France
设计者 / By : Territoires (project representative), Gelin & Lafon, Charles-Henri Tachon, Géosystème, Safege, CAEI, Boutté, Pascaline Minella, Nicolas Waltefaugle
业主 / For : Ardèche General Council
2.3 M€ ex. VAT

2013

诺尔热河发源地与湿地草原
WET MEADOWS AND SOURCE OF THE RIVER NORGES (P) pp.114-119
Norges-la-Ville, France
设计者 / By : Territoires (project representative)
业主 / For : Norges-la-Ville Town Council
310 000 € ex. VAT

2009

CONSOLATION VALLEY (C)
Val de Consolation, France
Award winners (abandoned projet)

2008

UPPER SEINE CANAL BIKE PATH (P)
France

2001

HAUT-JURA NATURAL PARK HOUSE (C)
Lajoux, France

1998

游艇码头 MARINA (P) pp.120-123
Vouglans, France
设计者 / By : Territoires (project representative), AAHJ architecte
业主 / For : Jura General Council
229 000 € ex. VAT

WATER SPORTS CENTRE (C)
Gérardmer, France

interventions-distinctions

活动与得奖记录

展览

- 《流动的城市》（与Michel Corajoud共同展出），国际设计双年展，法国圣艾蒂安，2010年
- 《巴黎，大都会与发展计划》，阿森纳展览馆永久收藏，法国巴黎，2011年
- 《转变的景观，改变土地的15个方案》，各地CAUE/FFP组织之巡回展，法国，2011-2013年
- 《节庆日》，梅林剧场论坛，瑞士梅林，2012年
- 《弗朗什-孔泰地区的现代与当代建筑》，建筑之家，法国弗朗什-孔泰地区，2013年
- 《2014 布拉格景观节》，建筑中心，捷克布拉格，2014年

讲座

- 《大地景观事务所（TERRITOIRES）的环境研究方法》，Ajena能源与环境协会，法国弗朗什-孔泰地区，2008年
- 斯特拉斯堡国立应用科学学院，法国斯特拉斯堡，2009年
- 《城市规划师郁景观设计师的合作关系》（与Nicolas Michelin共同演说），日内瓦景观、工程与建筑高等学院，瑞士日内瓦，2010年
- 洛桑联邦理工学院，瑞士洛桑，2012年
- 《波尔多尼尔城堡街区整治》（与Nicolas Michelin共同演说），萨沃伊建筑之家，法国萨沃伊，2012年
- 《城市小训示》，阿森纳展览馆，法国巴黎，2012年
- 《生态小区，建设与生活》，梅林剧场论坛，瑞士梅林，2012年
- 《花园与它的设计者》，贝桑松植物园，法国贝桑松，2013年

设计课

- 景观、工程与建筑高等学院，瑞士日内瓦，2005年起
- 蒲鲁东中学，法国贝桑松，2008年起
- 《生态大地》工作室，法国政府生态与可持续发展部，2011年8月
- 国立高等建筑学院，法国南希，2012年
- 国立高等美术学院，法国贝桑松，2013年

得奖

- 由Le Moniteur杂志颁发的城市空间整治奖，获奖方案：2004年葛雷的椴树广场，2010年塞尔芒吉的大草原
- 自从2002年以来多次获得由CAUE（建筑、城市与环境咨询委员会）颁发的建筑与城市空间整治奖项，获奖方案包括：沙勒泽尔的喷泉、葛雷的椴树广场、塞尔芒吉的大草原、富热罗莱的樱桃博物馆......

Exhibitions

- "La Ville Mobile / The Mobile City" (with Michel Corajoud), Saint-Étienne International Design Biennial, France, 2010
- "Paris, la métropole et ses projets / Paris, the Metropolis and its Projects", permanent collection at the Pavillon de l'Arsenal in Paris, France, 2011
- "Paysages transformés, 15 projets qui changent le territoire / Regenerated Landscapes, 15 projects that have changed the land", CAUE / FFP network touring exhibition, France, 2011-2013
- "Jour de fête / Gala Day", Meyrin Forum Theatre, Switzerland, 2012
- "Architecture Moderne et Contemporaine en Franche-Comté / Modern and Contemporary Architecture in Franche-Comté", Maison de l'Architecture in Franche-Comté, France, 2013
- "Landscape Festival Prague 2014", Architecture Centre, Czech Republic, 2014

Conferences

- "L'approche environnementale selon TERRITOIRES / The environmental approach according to TERRITOIRES", Ajena Franche-Comté, France, 2008
- INSA (National Institute of Applied Sciences) of Strasbourg, France, 2009
- "Les relations entre un urbaniste et un paysagiste / The relationship between a planner and a landscape architect" (with Nicolas Michelin), HEPIA Geneva, Switzerland, 2010
- Swiss Federal Institute of Technology in Lausanne, Switzerland, 2012
- " La Bastide Niel à Bordeaux / The Bastide Niel in Bordeaux" (with Nicolas Michelin), Maison de l'Architecture in Savoie, France, 2012
- "Petites leçons de ville / Small Lessons about Town", Pavillon de l'Arsenal, France, 2012
- "Construire et vivre dans un écoquartier / Building and Living in an Eco-Neighbourhood", Meyrin Forum Theatre, Switzerland, 2012
- "Le jardin et son concepteur / The Garden and its Designer", Besançon Botanical Garden, France, 2013

Workshops

- HEPIA (Architecture, Engineering and Landscaping University) Geneva, Switzerland, since 2005
- Proudhon College, Besançon, France, since 2008
- Ecological Land, Ministry of Ecology and Sustainable Development, August 2011
- Nancy National Institute of Architecture, France, 2012
- Besançon National Institute of Fine Art, France, 2013

Awards

- Moniteur Urban Planning Prize 2004 (Tilleuls Square, Gray) and 2010 (Large Meadow, Sermange)
- CAUE Architecture and Urban Planning Trophies and Awards since 2002, notably for the Chalezeule Fountain, Tilleuls Square, the Large Meadow in Sermange and the Cherry Museum in Fougerolles...

bibliography

出版

大地景观事务所（TERRITOIRE）的设计方案经常刊登于各类专业刊物，特别是以下杂志：

AMC, Anthos, Archdaily, Architectura et biznes, ARQA, Ecologik, Europaconcorsi, Gooood, Inhabitat, Landezine, Landscape Architecture, Le Moniteur, Paysage Actualités, Traits Urbains...

事务所的作品也在诸多建筑领域的国际性图书中被介绍，其中包括：

"In Touch" (Ed. Blauwdruk, 2011), "L'année du paysage en France" (Ed. FFP, 2011), "Paysages urbains : une France intime" (Pascal Dutertre, Ed. Le Moniteur, 2012), "Urban Park Landscape" (Ed. Chloe Fang, 2012), "Liquid Landscapes" (Ed. European Biennial of Landscape Architecture, 2012), "Design of Landscape" (Ed. Think Archit, 2012), "Global Architecture Impression II" (Ed. Saihan, 2012), "Territoires Économiques" (Ed. Ministère de l'Écologie, du Développement Durable et de l'Énergie, 2012), "Paysages transformés » (Ed. FFP, 2013), « Green City Spaces » (Ed. Braun, 2013), "Expression Paysagère 2" (Ed. ICI Interface, 2013), "Landscape Records: Community Gardens" (Ed; Profession Design Press, 2014)

TERRITOIRES work is published on a regular basis in specialized press, including:

AMC, Anthos, Archdaily, Architectura et biznes, ARQA, Ecologik, Europaconcorsi, Gooood, Inhabitat, Landezine, Landscape Architecture, Le Moniteur, Paysage Actualités, Traits Urbains...

TERRITOIRES projects are also presented in international Architecture books, such as:

"In Touch" (Ed. Blauwdruk, 2011), "L'année du paysage en France" (Ed. FFP, 2011), "Paysages urbains : une France intime" (Pascal Dutertre, Ed. Le Moniteur, 2012), "Urban Park Landscape" (Ed. Chloe Fang, 2012), "Liquid Landscapes" (Ed. European Biennial of Landscape Architecture, 2012), "Design of Landscape" (Ed. Think Archit, 2012), "Global Architecture Impression II" (Ed. Saihan, 2012), "Territoires Économiques" (Ed. Ministère de l'Écologie, du Développement Durable et de l'Énergie, 2012), "Paysages transformés » (Ed. FFP, 2013), « Green City Spaces » (Ed. Braun, 2013), "Expression Paysagère 2" (Ed. ICI Interface, 2013), "Landscape Records: Community Gardens" (Ed; Profession Design Press, 2014)

clients-partners

客户与合作者

AB Ingénieurs s.a., Addict Architecture, AEV Ile de France, Agence Bruno Fortier, ANMA, Agence ON, Agence Rudy Ricciotti Architecte, Air Architecte, Amiot-Lombard, Arcadis,AREP, Artélia, Atelier 981, Atelier Thibault Terrible, Berim, BG Ingénieurs conseils, Caen Presqu'île, CA Pays de Meaux, CA Pays de Montbéliard, CNF Clairefontaine, Charles-Henri Tachon Architecte, CU de Bordeaux, CU de Dijon, Commune d'Ixelles (Belgique), Conix, CG de l'Ardèche, CG des Ardennes, CM-CIC, D'Ascia Architecte, Damien Cabiron, Diagram, Dietman Feichtinger, DRAC de Franche-Comté, Ducks Sceno, Echologos, Elior, Emmanuel Gérard, EPA Marne, EPAMSA, Fabeck Architectes, Faire la Ville, Farra Zoumboulakis, C+S Fenon, Feraru, Christophe Gaudard, Geos, Ignacio Prego, Inexia, Ingerop, Integral 4, Integral Ruedi Baur, La Fabrique Urbaine, Le Point Lumineux, Light Cible, Meier & Associés, Metra & Associés, Pascaline Minella, Ministère de l'Écologie et du Développement Durable, Yves Morel, MVRDV, Oasiis, OFIS, OGI, Omnitech, OPAC de Saône-et-Loire, Osterle, Pro Développement, Quirot-Vichard, RFF, RFR, Safege, SEDD, SEMAVIP, SERM, SEURA, SNCF, SOCAD, Société du Grand Paris, Soderef, Solorem, SPLA, Studio Vicarini, Systra, TOA, Trans Faire, Transitec, Tribu, TVK, UMSF, Vanguard, Versailles Habitat, Ville d'Asnières-sur-Seine, Ville de Besançon, Ville de Béthune, Ville de Chalon-sur-Saône, Ville de Genève (Suisse), Ville de Gray, Ville de Grenoble, Ville de Madrid (Espagne), Ville de Meyrin (Suisse), Ville de Neuchâtel (Suisse), Ville de Nice, Ville de Paris, Ville de Sceaux, Ville de Strasbourg, Vincent Parreira Architecte, VNF...

AB Ingénieurs s.a., Addict Architecture, AEV Ile de France, Agence Bruno Fortier, ANMA, Agence ON, Agence Rudy Ricciotti Architecte, Air Architecte, Amiot-Lombard, Arcadis,AREP, Artélia, Atelier 981, Atelier Thibault Terrible, Berim, BG Ingénieurs conseils, Caen Presqu'île, CA Pays de Meaux, CA Pays de Montbéliard, CNF Clairefontaine, Charles-Henri Tachon Architecte, CU de Bordeaux, CU de Dijon, Commune d'Ixelles (Belgique), Conix, CG de l'Ardèche, CG des Ardennes, CM-CIC, D'Ascia Architecte, Damien Cabiron, Diagram, Dietman Feichtinger, DRAC de Franche-Comté, Ducks Sceno, Echologos, Elior, Emmanuel Gérard, EPA Marne, EPAMSA, Fabeck Architectes, Faire la Ville, Farra Zoumboulakis, C+S Fenon, Feraru, Christophe Gaudard, Geos, Ignacio Prego, Inexia, Ingerop, Integral 4, Integral Ruedi Baur, La Fabrique Urbaine, Le Point Lumineux, Light Cible, Meier & Associés, Metra & Associés, Pascaline Minella, Ministère de l'Écologie et du Développement Durable, Yves Morel, MVRDV, Oasiis, OFIS, OGI, Omnitech, OPAC de Saône-et-Loire, Osterle, Pro Développement, Quirot-Vichard, RFF, RFR, Safege, SEDD, SEMAVIP, SERM, SEURA, SNCF, SOCAD, Société du Grand Paris, Soderef, Solorem, SPLA, Studio Vicarini, Systra, TOA, Trans Faire, Transitec, Tribu, TVK, UMSF, Vanguard, Versailles Habitat, Ville d'Asnières-sur-Seine, Ville de Besançon, Ville de Béthune, Ville de Chalon-sur-Saône, Ville de Genève (Suisse), Ville de Gray, Ville de Grenoble, Ville de Madrid (Espagne), Ville de Meyrin (Suisse), Ville de Neuchâtel (Suisse), Ville de Nice, Ville de Paris, Ville de Sceaux, Ville de Strasbourg, Vincent Parreira Architecte, VNF...

团队 team

目前团队

Philippe Convercey, 项目总监
Franck Mathé, 技术总监
Étienne Voiriot, 执行总监
Pierre-Emmanuel Parisot, 建筑师
Linda Seyve, 项目负责人, 景观设计师
Delphine Dubreuil, 行政助理
Carole Bertrand, 媒体宣传
Gaël Gouhier, 景观绘图员

从前工作伙伴

Julian Aragundi Lamazza, Karin Arén, Florian Aubry, Alexandre Audonnet, Aurore Ballot, Anaïs Becker, Esia Belbachir, Clément Bertholet, Christophe Bonjour, Stéphanie Canada, Shi Cao, Thomas Cattin, Étienne Charrière, Bérengère Clerc, Carine Cucherousset, Chloé Davois, Adrien Do Nascimiento, Sylvain Dujeancourt, Samuel Enjolras, Blanca Franco, Ludivine Gérardin, Giulio Giorgi, Frédéric Goujet, Mathilde Harmand, Guillaume Haton, Gaëlle Hermabessière, Brice Huot Soudain, Eliott Jeanningros, Emma Journet, Jean Baptiste Leduc, Félicien Lesec, Romane Lhommée, Julie Lienard, Virginie Lolivier, Maëlle Lorinier, Aline Maire, David Maire, Thibaut Martini, Guillaume Mougenot, Amandine Parret Rebecca Perret, Sébastien Perret, Charles Petit-Imbert, Elvina Piard, Antonin Picard, Daniel Planella i Oriol, Marta Puig i Bosch, Julien Rechautier, Jérémy Roussel, Zoé Rundstadler-Schneider, Cédric Sattler, Florian Sommer, Ninon Szejman, Chloé Vichard.

Current team

Philippe Convercey, Projects Design Manager
Franck Mathé, Technical Manager
Étienne Voiriot, Operations Manager
Pierre-Emmanuel Parisot, Architect
Linda Seyve, Chef de projet, Landscape Architect
Delphine Dubreuil, Managing Assistant
Carole Bertrand, Communication & Medias
Gaël Gouhier, Landscape Draughtsman

Former collaborators

Julian Aragundi Lamazza, Karin Arén, Florian Aubry, Alexandre Audonnet, Aurore Ballot, Anaïs Becker, Esia Belbachir, Clément Bertholet, Christophe Bonjour, Stéphanie Canada, Shi Cao, Thomas Cattin, Étienne Charrière, Bérengère Clerc, Carine Cucherousset, Chloé Davois, Adrien Do Nascimiento, Sylvain Dujeancourt, Samuel Enjolras, Blanca Franco, Ludivine Gérardin, Giulio Giorgi, Frédéric Goujet, Mathilde Harmand, Guillaume Haton, Gaëlle Hermabessière, Brice Huot Soudain, Eliott Jeanningros, Emma Journet, Jean Baptiste Leduc, Félicien Lesec, Romane Lhommée, Julie Lienard, Virginie Lolivier, Maëlle Lorinier, Aline Maire, David Maire, Thibaut Martini, Guillaume Mougenot, Amandine Parret Rebecca Perret, Sébastien Perret, Charles Petit-Imbert, Elvina Piard, Antonin Picard, Daniel Planella i Oriol, Marta Puig i Bosch, Julien Rechautier, Jérémy Roussel, Zoé Rundstadler-Schneider, Cédric Sattler, Florian Sommer, Ninon Szejman, Chloé Vichard.

致谢 acknowledgments

特别感谢Nicolas Waltefaugle (nwphoto.fr) 的协助，他多年来以其摄影和友谊伴随大地景观事务所（TERRITOIRES）的成长，也感谢 Rudy Ricciotti 为本书撰写的短序。

感谢亦西文化出版社（ICI Interface / ICI Consultants）让我们的作品能够在《绿色观点Green Vision》这个系列呈现。

感谢二十年来不断支持大地景观事务所的所有的工作伙伴、版面设计师、专业顾问、合作者、联合承办、客户、记者、出版社，以及我们的朋友与家人。由于他们，本书中所呈现的项目与工作才能开花结果。

最后，我们要感谢提供本书图面资料的事务所、绘图者和摄影师，谢谢他们同意我们在本书中使用他们的作品。

Special thanks to Nicolas Waltefaugle (nwphoto.fr) whose photographs and friendship have accompanied TERRITOIRES' work for so many years, and to Rudy Ricciotti, for his preface.

We would also like to thank Editions ICI Interface (ICI Consultants) for having invited us to join this collection.

Our grateful thanks to every collaborator, graphic designer, consultant, partner, joint contractor, client, journalist, publisher, and to our friends and families, who have been supportive of TERRITOIRES for over 20 years. Without them, the work presented in this book would not have been possible.

Finally, we wish to thank the agencies, designers and photographers who allowed us to share their work in this book.

credits

版权说明

贝桑松Besançon – 巴黎Paris – 日内瓦Geneva

事务所总部 Headquarters :
22 rue Mégevand
25000 Besançon France

T +33 (0)3 81 82 06 66
info@territoirespaysagistes.com

www.territoirespaysagistes.com
www.territoireslandscapearchitects.com

社交网络 Social Medias

请于以下社交网络与大地景观事务所（TERRITOIRES）互动
Join TERRITOIRES on social networks
facebook.com/territoireslandscapearchitects
territoirespaysagistes.tumblr.com
linkedin.com/in/territoires

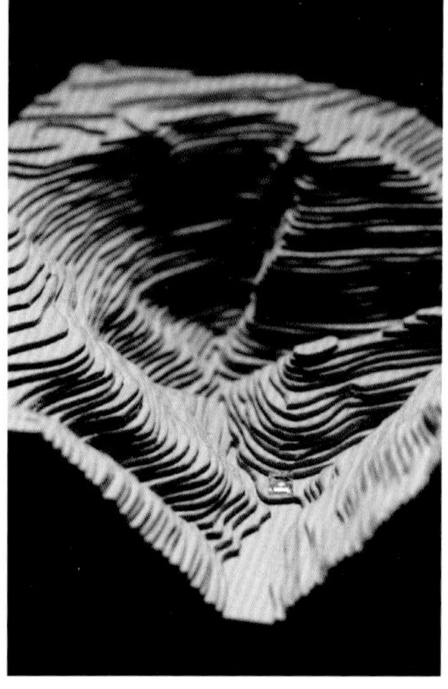

图书在版编目（CIP）数据

实践景观：大地景观事务所设计作品专辑TERRITOIRES / 法国亦西文化编；简嘉玲译. -- 沈阳：辽宁科学技术出版社，2014.8
（绿色观点. 景观设计师作品系列）
ISBN 978-7-5381-8760-1

Ⅰ. ①实… Ⅱ. ①法… ②简… Ⅲ. ①景观设计－作品集－法国－现代 Ⅳ. ①TU986.2

中国版本图书馆CIP数据核字(2014)第176639号

出版发行：辽宁科学技术出版社
（地址：沈阳市和平区十一纬路29号 邮编：110003）
印 刷 者：利丰雅高印刷（深圳）有限公司
经 销 者：各地新华书店
幅面尺寸：210mm×230mm
印 张：8.5
字 数：100千字
印 数：1～1500
出版时间：2014年 8 月第 1 版
印刷时间：2014年 8 月第 1 次印刷
责任编辑：陈慈良 宋丹丹
封面设计：维建·诺黑
版式设计：维建·诺黑
责任校对：周 文
书 号：ISBN 978-7-5381-8760-1
定 价：88.00元

联系电话：024-23284360
邮购热线：024-23284502
E-mail: lnkjc@126.com
http://www.lnkj.com.cn
本书网址：www.lnkj.cn/uri.sh/8760